Physical testing of textiles

Physical testing
of textiles

B P Saville

The Textile Institute

CRC Press
Boca Raton Boston New York Washington, DC

WOODHEAD PUBLISHING LIMITED

Oxford   Cambridge   New Delhi

Published by Woodhead Publishing Limited in association with The Textile Institute
Woodhead Publishing Limited, Abington Hall, Granta Park, Great Abington
Cambridge CB21 6AH, UK
www.woodheadpublishing.com

Woodhead Publishing India Private Limited, G-2, Vardaan House, 7/28 Ansari Road,
Daryaganj, New Delhi – 110002, India

Published in North America by CRC Press LLC, 6000 Broken Sound Parkway, NW
Suite 300, Boca Raton FL 33487, USA

First published 1999, Woodhead Publishing Limited and CRC Press LLC
Reprinted 2000, 2002, 2003, 2007, 2009
© 1999, Woodhead Publishing Limited
The author has asserted his moral rights.

British Library Cataloguing in Publication Data
A catalogue record for this book is available from the British Library.

Library of Congress Cataloging in Publication Data
A catalog record for this book is available from the Library of Congress.

Woodhead Publishing ISBN 978-1-85573-367-1 (book)
Woodhead Publishing ISBN 978-1-84569-015-1 (e-book)
CRC Press ISBN 978-0-8493-0568-9
CRC Press order number: WP0568

Printed by TJI Digital, Padstow, Cornwall, UK

Contents

Preface

This book arose out of a need, when teaching textile testing at Huddersfield University, for a modern volume on the subject to which students could be referred. The approach to the subject and the topics covered are ones that have been developed over the years in the textile department at Huddersfield. This institution has had, until now, a long history of teaching part-time students on day-release from their employment in the local textile industry. Because of this the testing section developed close links with the industry and thus became aware of the problems and requirements in this area. Therefore the content of the courses which were taught was developed over the years by a succession of workers in this area. This book is then a distillation of this collective wisdom and hopefully a further development of it. In particular I would like to mention my former colleagues Carol Ramsden, Philip Townhill and Christine Wilkinson, all of whom could have written this book but declined the chore.

The book is aimed both at students intending to join the textile industry and also at personnel who are already employed there in the areas of testing and quality assurance. Because modern industry produces goods, be they garments, fabrics, yarns or fibres, to specifications, the emphasis throughout this book is on standard and reproducible tests. The reason for this is that in order to specify a property it is also necessary at the same time to specify the method of test. Many of the tests carried out on textile materials are intended to measure the same property but because of their different approach or the type of equipment used can give different results.

It is important when carrying out any tests to consult an up-to-date copy of the relevant standard. This is because the actual standard contains a wealth of detail which it is not possible to cover in a book of this nature and furthermore national and international standards are constantly being changed and updated.

B P Saville

Acknowledgements

The author would like to thank Prof. J W S Hearle for reading through the manuscript and making many helpful suggestions.

Introduction

1.1 Reasons for textile testing

The testing of textile products is an expensive business. A laboratory has to be set up and furnished with a range of test equipment. Trained operatives have to be employed whose salaries have to be paid throughout the year, not just when results are required. Moreover all these costs are non-productive and therefore add to the final cost of the product. Therefore it is important that testing is not undertaken without adding some benefit to the final product.

There are a number of points in the production cycle where testing may be carried out to improve the product or to prevent sub-standard merchandise progressing further in the cycle.

1.1.1 Checking raw materials

The production cycle as far as testing is concerned starts with the delivery of raw material. If the material is incorrect or sub-standard then it is impossible to produce the required quality of final product.

The textile industry consists of a number of separate processes such as natural fibre production, man-made fibre extrusion, wool scouring, yarn spinning, weaving, dyeing and finishing, knitting, garment manufacture and production of household and technical products. These processes are very often carried out in separate establishments, therefore what is considered to be a raw material depends on the stage in processing at which the testing takes place. It can be either the raw fibre for a spinner, the yarn for a weaver or the finished fabric for a garment maker. The incoming material is checked for the required properties so that unsuitable material can be rejected or appropriate adjustments made to the production conditions. The standards that the raw material has to meet must be set at a realistic level. If the standards are set too high then material will be rejected that is good enough for the end use, and if they are set too low then large amounts of inferior material will go forward into production.

1.1.2 Monitoring production

Production monitoring, which involves testing samples taken from the production line, is known as quality control. Its aim is to maintain, within known tolerances, certain specified properties of the product at the level at which they have been set. A quality product for these purposes is defined as one whose properties meets or exceeds the set specifications.

Besides the need to carry out the tests correctly, successful monitoring of production also requires the careful design of appropriate sampling procedures and the use of statistical analysis to make sense of the results.

1.1.3 Assessing the final product

In this process the bulk production is examined before delivery to the customer to see if it meets the specifications. By its nature this takes place after the material has been produced. It is therefore too late to alter the production conditions. In some cases selected samples are tested and in other cases all the material is checked and steps taken to rectify faults. For instance some qualities of fabric are inspected for faulty places which are then mended by skilled operatives; this is a normal part of the process and the material would be dispatched as first quality.

1.1.4 Investigation of faulty material

If faulty material is discovered either at final inspection or through a customer complaint it is important that the cause is isolated. This enables steps to be taken to eliminate faulty production in future and so provide a better quality product. Investigations of faults can also involve the determination of which party is responsible for faulty material in the case of a dispute between a supplier and a user, especially where processes such as finishing have been undertaken by outside companies. Work of this nature is often contracted out to independent laboratories who are then able to give an unbiased opinion.

1.1.5 Product development and research

In the textile industry technology is changing all the time, bringing modified materials or different methods of production. Before any modified product reaches the market place it is necessary to test the material to check that the properties have been improved or have not been degraded by faster production methods. In this way an improved product or a lower-cost product with the same properties can be provided for the customer. A large organisation will often have a separate department to carry out research and development; otherwise it is part of the normal duties of the testing department.

1.2   Standardisation of testing

When a textile material is tested certain things are expected from the results. Some of these are explicit but other requirements are implicit. The explicit requirements from the results are either that they will give an indication of how the material will perform in service or that they will show that it meets its specification.

The implicit requirement from a test is that it is reproducible, that is if the same material is tested either at another time, or by another operator or in a different laboratory the same values will be obtained. In other words the test measures some 'true' or correct value of the property being assessed. If the test results vary from laboratory to laboratory then the test is not measuring anything real and it is pointless carrying it out. However, the values that are obtained from testing textile materials are not expected to be exactly the same, so that appropriate statistical criteria should be applied to the results to see whether they fall within the accepted spread of values.

The lack of reproducibility of test results can be due to a number of causes.

1.2.1 Variation in the material

Most textile materials are variable, natural fibres having the most variation in their properties. The variation decreases as the production progresses from fibres to yarns to fabrics, since the assembly of small variable units into larger units helps to smooth out the variation in properties. The problem of variable material can be dealt with by the proper selection of representative samples and the use of suitable statistical methods to analyse the results.

1.2.2 Variation caused by the test method

It is important that any variations due to the test itself are kept to the minimum. Variability from this source can be due to a number of causes:

1   The influence of the operator on the test results. This can be due to differences in adherence to the test procedures, care in the mounting of specimens, precision in the adjustment of the machine such as the zero setting and in the taking of readings.
2   The influence of specimen size on the test results, for instance the effect of specimen length on measured strength.

3   The temperature and humidity conditions under which the test is carried out. A number of fibres such as wool, viscose and cotton change their properties as the atmospheric moisture content changes.
4   The type and make of equipment used in the test. For instance pilling tests can be carried out using a pilling box or on the Martindale abrasion machine. The results from these two tests are not necessarily comparable.
5   The conditions under which the test is carried out such as the speed, pressure or duration of any of the factors.

It is therefore necessary even within a single organisation to lay down test procedures that minimise operator variability and set the conditions of test and the dimensions of the specimen. Very often in such cases, factors such as temperature, humidity and make of equipment are determined by what is available.

However, when material is bought or sold outside the factory there are then two parties to the transaction, both of whom may wish to test the material. It therefore becomes important in such cases that they both get the same result from testing the same material. Otherwise disputes would arise which could not be resolved because each party was essentially testing a different property.

This requires that any test procedures used by more than one organisation have to be more carefully specified, including, for instance, the temperature and humidity levels at which the test takes place. The details in the procedure have to be sufficient so that equipment from different manufacturers will produce the same results as one another. This need for standard written test methods leads to the setting up of national standards for test procedures so making easier the buying and selling of textiles within that country. Even so certain large organisations, such as IWS or Marks and Spencer, have produced their own test procedures to which suppliers have to conform if they wish to carry the woolmark label or to sell to Marks and Spencer.

Most countries have their own standards organisations for example: BS (Britain), ASTM (USA) and DIN (Germany) standards. The same arguments that are used to justify national standards can also be applied to the need for international standards to assist world-wide trade, hence the existence of International Organization for Standardization (ISO) test methods and, within the European Union, the drive to European standards.

1.3   Sampling

It is not possible or desirable to test all the raw material or all the final output from a production process because of time and cost constraints. Also many tests are destructive so that there would not be any material left after it had been tested. Because of this, representative samples of the material are tested. The amount of material that is actually tested can represent a very small proportion of the total output. It is therefore important that this small sample should be truly representative of the whole of the material. For instance if the test for cotton fibre length is considered, this requires a 20 mg sample which may have been taken from a bale weighing 250 kg. The sample represents only about one eleven-millionth of the bulk but the quality of the whole bale is judged on the results from it.

The aim of sampling is to produce an unbiased sample in which the proportions of, for instance, the different fibre lengths in the sample are the same as those in the bulk. Or to put it another way, each fibre in the bale should have an equal chance of being chosen for the sample [1].

1.3.1 Terms used in sampling

Several of the terms used in sampling have different meanings depending on whether wool or cotton, yarn or fibre is being sampled. This is due to the different representative organisations which have historically grown around each industry. The appropriate standard should always be consulted [1–4]:

- **Consignment**: this is the quantity of material delivered at the same time. Each consignment may consist of one or several lots.
- **Test lot or batch**: this consists of all the containers of a textile material of one defined type and quality, delivered to one customer according to one dispatch note. The material is presumed to be uniform so that this is the whole of the material whose properties are to be characterised by one set of tests. It can be considered to be equivalent to the statistical population.
- **Laboratory sample**: this is the material that will be used as a basis for carrying out the measurement in the laboratory. This is derived by appropriate random sampling methods from the test lot.
- **Test specimen**: this is the one that is actually used for the individual measurement and is derived from the laboratory sample. Normally, measurements are made from several test specimens.
- **Package**: elementary units (which can be unwound) within each container in the consignment. They might be bump top, hanks, skeins, bobbins, cones or other support on to which have been wound tow, top, sliver, roving or yarn.
- **Container or case**: a shipping unit identified on the dispatch note, usually a carton, box, bale or other container which may or may not contain packages.

1.3.2 Fibre sampling from bulk

Zoning

Zoning is a method that is used for selecting samples from raw cotton or wool or other loose fibre where the properties may vary considerably from place to place. A handful of fibres is taken at random from each of at least 40 widely spaced places (zones) throughout the bulk of the consignment and is treated as follows. Each handful is divided into two parts and one half of it is discarded at random; the retained half is again divided into two and half of that discarded. This process is repeated until about n/x fibres remain in the handful (where n is the total number of fibres required in the sample and x is the number of original handfuls). Each handful is treated in a similar manner and the fibres that remain are placed together to give a correctly sized test sample containing n fibres. The method is shown diagrammatically in Fig. 1.1. It is important that the whole of the final sample is tested.

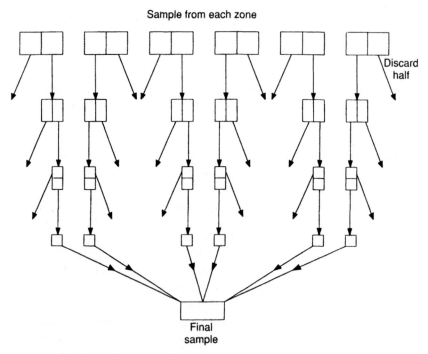

Sample from each zone

Discard half

Final sample

1.1 Sampling by zoning.

Core sampling

Core sampling is a technique that is used for assessing the proportion of grease, vegetable matter and moisture in samples taken from unopened bales of raw wool. A tube with a sharpened tip is forced into the bale and a core of wool is withdrawn. The technique was first developed as core boring in which the tube was rotated by a portable electric drill. The method was then developed further [5] to enable the cores to be cut by pressing the tube into the bale manually. This enables samples to be taken in areas remote from sources of power.

The tubes for manual coring are 600 mm long so that they can penetrate halfway into the bale, the whole bale being sampled by coring from both ends. A detachable cutting tip is used whose internal diameter is slightly smaller than that of the tube so that the cores will slide easily up the inside of the tube. The difference in diameter also helps retain the cores in the tube as it is withdrawn. To collect the sample the tube is entered in the direction of compression of the bale so that it is perpendicular to the layers of fleeces. A number of different sizes of nominal tube diameter are in use, 14, 15 and 18 mm being the most common the weight of core extracted varying accordingly. The number of cores extracted is determined according to a sampling schedule [6] and the cores are combined to give the required weight of sample. As the cores are removed they are placed immediately in an air-tight container to prevent any loss of moisture from them. The weight of the bale at the time of coring is recorded in order to calculate its total moisture content.

The method has been further developed to allow hydraulic coring by machine in warehouses where large numbers of bales are dealt with. Such machines compress the bale to 60% of its original length so as to allow the use of a tube which is long enough to core the full length of the bale.

1.3.3 Fibre sampling from combed slivers, rovings and yarn

One of the main difficulties in sampling fibres is that of obtaining a sample that is not biased. This is because unless special precautions are taken, the longer fibres in the material being sampled are more likely to be selected by the sampling procedures, leading to a length-biased sample. This is particularly likely to happen in sampling material such as sliver or yarn where the fibres are approximately parallel. Strictly speaking, it is the fibre extent as defined in Fig. 1.2 rather than the fibre length as such which determines the likelihood of selection. The obvious area where length bias must be avoided is in the measurement of fibre length, but any bias can also have effects when other properties such as fineness and strength are being mea-

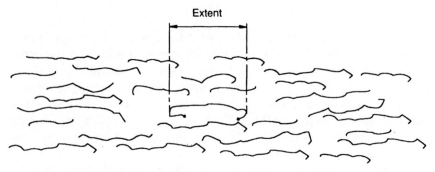

1.2 The meaning of extent.

sured since these properties often vary with the fibre length. There are two ways of dealing with this problem:

1 Prepare a numerical sample (unbiased sample).
2 Prepare a length-biased sample in such a way that the bias can be allowed for in any calculation.

Numerical sample

In a numerical sample the percentage by number of fibres in each length group should be the same in the sample as it is in the bulk. In Fig. 1.3, A and B represent two planes separated by a short distance in a sample consisting of parallel fibres. If all the fibres whose left-hand ends (shown as solid circles) lay between A and B were selected by some means they would constitute a numerical sample. The truth of this can be seen from the fact that if all the fibres that start to the left of A were removed then it would not alter the marked fibres. Similarly another pair of planes could be imagined to the right of B whose composition would be unaffected by the removal of the fibres starting between A and B. Therefore the whole length of the sample could be divided into such short lengths and there would be no means of distinguishing one length from another, provided the fibres are uniformly distributed along the sliver. If the removal of one sample does not affect the composition of the remaining samples, then it can be considered to be a numerical sample and each segment is representative of the whole.

Length-biased sample

In a length-biased sample the percentage of fibres in any length group is proportional to the product of the length and the percentage of fibres of

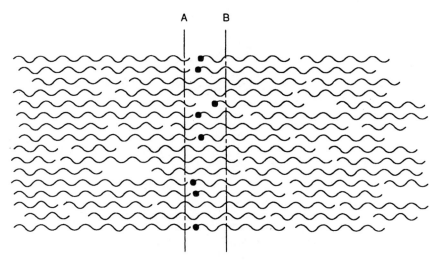

1.3   Selection of a numerical sample.

that length in the bulk. The removal of a length-biased sample changes the composition of the remaining material as a higher proportion of the longer fibres are removed from it.

If the lines A and B in Fig. 1.3 represent planes through the sliver then the chance of a fibre crossing these lines is proportional to its length. If, therefore, the fibres crossing this area are selected in some way then the longer fibres will be preferentially selected. This can be achieved by gripping the sample along a narrow line of contact and then combing away any loose fibres from either side of the grips, so leaving a sample as depicted in Fig. 1.4 which is length-biased. This type of sample is also known as a tuft sample and a similar method is used to prepare cotton fibres for length measurement by the fibrograph. Figure 1.5 shows the fibre length histogram and mean fibre length from both a numerical sample and a length-biased sample prepared from the same material [7].

By a similar line of reasoning if the sample is cut at the planes A and B the section between the planes will contain more pieces of the longer fibres because they are more likely to cross that section. If there are equal numbers of fibres in each length group, the total length of the group with the longest fibres will be greater than that of the other groups so that there will be a greater number of those fibres in the sample. Samples for the measurement of fibre diameter using the projection microscope are prepared in this manner by sectioning a bundle of fibres, thus giving a length-biased sample. The use of a length-biased sample is deliberate in this case so that the measured mean fibre diameter is then that of the total fibre length of the whole sample. If all the fibres in the sample are considered as being

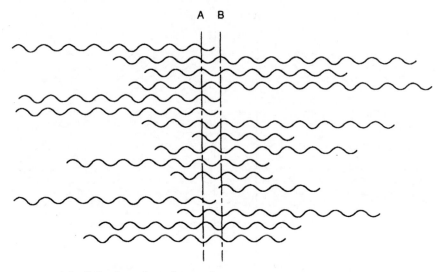

1.4   Selection of a tuft sample.

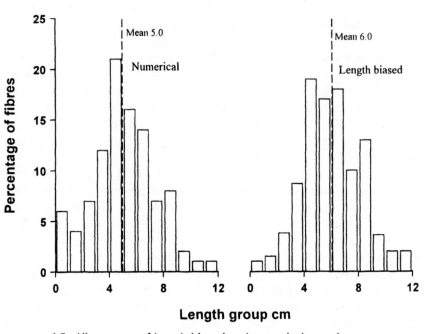

1.5   Histograms of length-biased and numerical samples.

joined end to end the mean fibre diameter is then the average thickness of that fibre.

Random draw method

This method is used for sampling card sliver, ball sliver and top. The sliver to be sampled is parted carefully by hand so that the end to be used has no broken or cut fibres. The sliver is placed over two velvet boards with the parted end near the front of the first board. The opposite end of the sliver is weighed down with a glass plate to stop it moving as shown in Fig. 1.6. A wide grip which is capable of holding individual fibres is then used to remove and discard a 2 mm fringe of fibres from the parted end. This procedure is repeated, removing and discarding 2 mm draws of fibre until a distance equal to that of the longest fibre in the sliver has been removed. The sliver end has now been 'normalised' and any of the succeeding draws can be used to make up a sample as they will be representative of all fibre lengths. This is because they represent a numerical sample as described above where all the fibres with ends between two lines are taken as the sample. When any measurements are made on such a sample all the fibres must be measured.

Cut square method

This method is used for sampling the fibres in a yarn. A length of the yarn being tested is cut off and the end untwisted by hand. The end is laid on a small velvet board and covered with a glass plate. The untwisted end of the yarn is then cut about 5 mm from the edge of the plate as shown in Fig. 1.7. All the fibres that project in front of the glass plate are removed one by one with a pair of forceps and discarded. By doing this all the cut fibres are removed, leaving only fibres with their natural length. The glass plate is then moved back a few millimetres, exposing more fibre ends. These are then removed one by one and measured. When these have all been measured the plate is moved back again until a total of 50 fibres have been measured. In each case once the plate has been moved all projecting fibre ends must be removed and measured. The whole process is then repeated on fresh lengths of yarn chosen at random from the bulk, until sufficient fibres have been measured.

1.3.4 Yarn sampling

When selecting yarn for testing it is suggested [8] that ten packages are selected at random from the consignment. If the consignment contains more than five cases. five cases are selected at random from it. The test

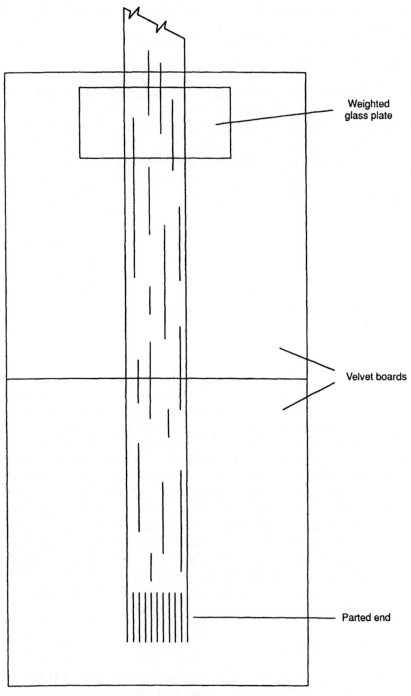

Weighted
glass plate

Velvet boards

Parted end

1.6   The random draw method.

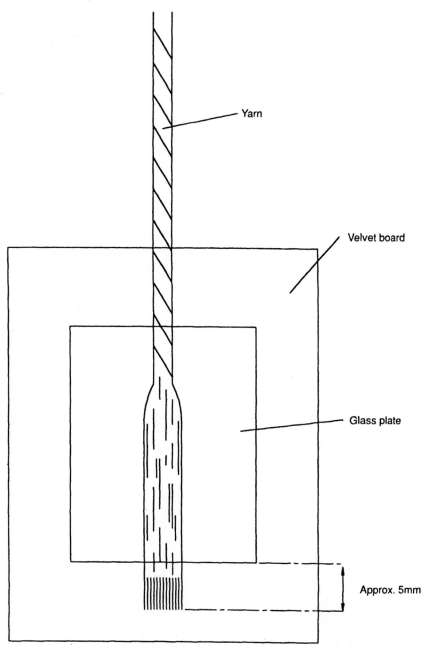

1.7   The cut square method.

sample then consists of two packages selected at random from each case. If the consignment contains less than five cases, ten packages are selected at random from all the cases with approximately equal numbers from each case. The appropriate number of tests are then carried out on each package.

1.3.5 Fabric sampling

When taking fabric samples from a roll of fabric certain rules must be observed. Fabric samples are always taken from the warp and weft separately as the properties in each direction generally differ. The warp direction should be marked on each sample before it is cut out. No two specimens should contain the same set of warp or weft threads. This is shown diagrammatically in Fig. 1.8 where the incorrect layout shows two warp samples which contain the same set of warp threads so that their properties will be very similar. In the correct layout each sample contains a different set of warp threads so that their properties are potentially different depending on the degree of uniformity of the fabric. As it is the warp direction in this case that is being tested the use of the same weft threads is not

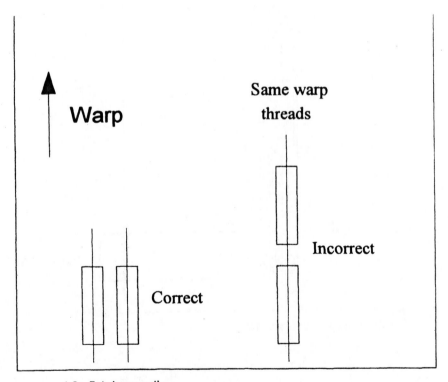

1.8   Fabric sampling.

important. Samples should not be taken from within 50 mm of the selvedge as the fabric properties can change at the edge and they are no longer representative of the bulk.

1.4   Measurement

The process of measurement can be defined as a quantitative comparison between a predefined standard and the object being measured. This definition shows that there are two parts to the measuring process: the comparison, which is the process that is usually thought of as measurement, and the predefined standard, which is the part that is easily overlooked. When an object is weighed in the laboratory on a single pan balance it gives a reading of the mass of the object and so the balance is the local standard. However, what is actually taking place is a comparison of the mass of the object with that of the international standard kilogram. The validity of the measurement relies on there being a clear link between the balance that is in use and the international standard. In other words the balance needs to be calibrated with standard masses that have themselves been calibrated against other masses that in turn have been calibrated against the international standard. This link needs to be documented at each calibration step as to when it was carried out and to what limits of accuracy it has been made so that the calibration of a given instrument can be traced back to the international standards. It is also important that this calibration is carried out at regular intervals as instrumental readings can change over time because of wear of mechanical parts and ageing of electronic circuits. Besides regular calibration being good laboratory practice it is specifically demanded by ISO 9000.

The actual process of measurement is always subject to errors which can be defined as the difference between the measured value and the 'true' value. However, the 'true' value of any parameter can never be known because the value can only be obtained through measurement and any measurement can only be an estimation of the value, subject to unknown errors.

The term **precision** as used by metrologists [9] means the same as repeatability. It is defined as the quality that characterises the ability of a measuring instrument to give the same value of the quantity measured. In other words it is an indication of how well identically performed measurements agree with each other. A measurement of a property may return a value of 2.9347, which because of the number of figures after the decimal point may impress with its precision. If the measurement is then immediately repeated by the operator on the same object a value of 2.8962 may be obtained, when it will be seen that the number of figures represents a spurious precision and that the actual precision is much less. The precision of any measurement can only be obtained by making a number of identical measurements and estimating the dispersion of the results about the mean. The standard deviation or coefficient of variation of a set of results is used as a measure of this. A single measurement is always of an unknown precision, although in general the precision of particular test procedures is known through repeated testing in a single laboratory.

However, a result may be very precise in that every time the measurement is made the same number is obtained but it may vary from the 'true' value due to systematic errors.

Accuracy may be defined as conformity with or nearness to the 'true' value of the quantity being measured. This can only be obtained by calibration of the measuring system against the appropriate standards at suitable intervals.

Sensitivity is defined as the least change in the measured quantity that will cause an observable change in the instrument reading. The sensitivity of a measuring instrument can be increased by amplifying the output or by using a magnifying lens to read the scale. Without an accompanying increase in the accuracy of the calibration and a reduction in sources of variation this may mean no more than an amplification of the errors as well.

1.4.1 Statistical terms

Most measurements in textile testing consist of a set of repeat measurements that have been made on a number of identical individuals constituting a sample taken from the bulk of the material. Certain statistical measures are used to describe the average of the results and their spread. A short guide to the terms employed is given below. For a more comprehensive explanation a textbook of statistics should be consulted [10–12].

Arithmetic mean or average

The arithmetic mean is the measure most commonly chosen to represent the central value of a sample. It is obtained by adding together the individual values of the variable x and dividing the sum by the number of individuals n. It is represented by the symbol $\bar{x}$:

$$\bar{x} = \frac{x_1 + x_2 + x_3 + \ldots + x_n}{n}$$

Standard deviation

The standard deviation is the most widely used measure of the dispersion or spread of results about the mean value. The symbol σ is used for the standard deviation of the universe (population) containing all the possible measurements that could be made of the variable in question. The symbol *s* is generally used for the estimated value of the standard deviation from a sample which has been taken from the universe

$$s = \sqrt{\left[\frac{\sum(x - \bar{x})^2}{n - 1}\right]}$$

The units that the standard deviation is measured in are the same as those of the mean.

Coefficient of variation

A standard deviation of 1 for a property that has a mean value of 10 is far more significant than a standard deviation of 1 for a property with a mean value of 100. Because of this the coefficient of variation (CV) is often used as a measure of dispersion: it is the standard deviation expressed as a percentage of the mean. Therefore in the above example the first result would have a CV of 10% and the second result would have a CV of only 1%.

$$CV = \frac{\text{standard deviation} \times 100}{\text{mean}} = \frac{s}{\bar{x}} \times 100\%$$

Standard error of the mean

The standard error of the mean is a measure of the reliability of the mean value obtained from a sample of a particular size. It is the standard deviation of the means that would be obtained if repeated samples of the given size were measured:

$$\text{Standard error of mean} = \frac{\sigma}{\sqrt{n}}$$

where σ is standard deviation of the parent universe. In the case of a sample the standard error of the mean has to be estimated by using the standard deviation of the sample *s* in place of σ.

The standard error can be used to place confidence limits on the mean that has been measured. For example there is a 95% probability that the population mean lies within $\pm(1.96 \times \text{standard error})$ of the measured mean value. This relationship only holds when the standard error has been calculated from the standard deviation of the parent universe σ or when the sample is large. For small samples where *s* has been used to calculate the standard error, the value of 1.96 should be replaced by the appropriate value of *t* obtained from statistical tables.

1.4.2 Determination of number of tests

In any test the number of individuals to be tested will depend on the variability of the material and the accuracy required from the measurement [1, 13]. If the material is repeatedly sampled at random and the test performed on n selected items each time, in 95% of cases the mean value which is calculated will be within $\pm(2C/\sqrt{n})\%$ of the population mean, where C is the coefficient of variation of the property being tested and n is the number of test specimens. The values of $(2C/\sqrt{n})\%$ are the confidence limits of error. For many standard tests the coefficient of variation is known approximately so that the number of tests necessary to achieve given confidence limits of error can then be calculated. For instance if the coefficient of variation for a yarn strength test is 10% and the number of tests carried out n is 5, there is a 95% chance that the mean value will lie within $\pm8.9\%$ of the population mean. If the number of tests is increased to 10, then there is the same chance that the mean will lie within $\pm6.3\%$ of the population mean, and if the number of tests is increased to 50, then it is likely that the mean will lie within $\pm2.8\%$ of the population mean.

The use of the coefficient of variation in the above formula assumes that the error, in the form of the standard deviation, is proportional to the mean value. For example in the above case of yarn strength if the mean value was 10N then the standard deviation would be 1N, whereas if the mean value was 100N then the standard deviation would be 10N. With some measurements the error is relatively independent of the magnitude of the mean. If this is the case then the actual standard deviation should be used instead of the coefficient of variation so that in 95% of cases the measured mean value will lie within $\pm2S/\sqrt{n}$ of the population mean, where S is the standard deviation.

1.4.3 Use of computers

The incorporation of computers and microprocessors has brought great changes to the instrumentation used for testing textiles. Their use falls into two main categories: recording and calculation of results and automation of the test procedure. Both of these uses may be found in the most advanced instruments.

Recording of results

In these applications the computer is usually connected via an analogue to digital converter to an existing instrument from where it collects the data that would previously have been written down on paper by the operator. The advantages of such an installation are as follows:

1   In the case of a graphical output the whole of the curve is recorded numerically so that results such as maxima, areas under the curve and slopes can be calculated directly without having to be read from a graph. This allows a more consistent measurement of features such as slopes which would previously have been measured by placing a rule on the graph by eye. However, it is important in such applications to be clear what criteria the computer is using to select turning points in the curve and at what point the slope is being measured. It is useful to have visual checks on these points in case the computer is making the wrong choice.

2   The ability to adjust the zero level for the instrument automatically. This can be done, for instance, by taking the quiescent output as being the zero level and subtracting this from all other readings.

3   The ability to perform all the intermediate calculations together with any statistical calculations in the case of multiple tests.

4   The ability to give a final neatly printed report which may be given directly to a customer.

It is important, however, to be aware of the fact that the precision of the basic instrument is unchanged and it depends on, among other things, the preparation and loading of the sample into the instrument by the operator and the setting of any instrumental parameters such as speed or range.

Automation of the test procedure

In such applications use is made of electronic processing power to control various aspects of the test rather than just to record the results. This means that steps such as setting ranges, speeds, tensions and zeroing the instrument can all be carried out without the intervention of an operator. The settings are usually derived from sample data entered at the keyboard. In the case of yarn-testing instruments the automation can be carried as far as loading the specimen. This enables the machinery to be presented with a number of yarn packages and left to carry out the required number of tests on each package.

The automation of steps in the test procedure enables an improvement to be made in the repeatability of test results owing to the reduction in operator intervention and a closer standardisation of the test conditions. The precision of the instrument is then dependent on the quality of the sensors and the correctness of the sample data given to the machine. The accuracy of the results is, however, still dependent on the calibration of the instrument. This is a point that is easily overlooked in instruments with digital outputs as the numbers have lost their immediate connection with the physical world. If the machine fails in some way but is still giving a numerical output, the figures may still be accepted as being correct.

To be generally acceptable automated instruments have to be able to carry out the test to the appropriate standard or have to be able to demonstrate identical results to those that have been obtained with the standard test method.

It is still possible even with advanced automation for results to be incorrect for such simple reasons as wrong identification of samples or failure to condition samples in the correct testing atmosphere.

1.4.4 Types of error

Errors fall into two types.

Bias or systematic error

With this type of error the measurements are consistently higher or lower than they should be. For instance if a balance is not zeroed before use then all readings taken from it will have the same small amount added to or subtracted from them. This type of error cannot be detected by any statistical examination of the readings. Systematic errors can only be eliminated by careful design of the tests, proper calibration and correct operation of the instruments.

Precision or random error

This type of error is present when repeated measurements of the same quantity give rise to differing values which are scattered at random around some central value. In such cases the error can be estimated by statistical methods.

1.4.5 Sources of error

Errors of both types can arise from a number of causes:

1 **Instrument reproducibility**: even when an instrument is correctly calibrated, mechanical defects can influence the readings unless they are taken in exactly the same fashion as the calibration values. Mechanical defects such as slackness, friction and backlash can cause measurements to vary. These effects can depend on the direction that the mechanism is moving so that the error may be different when the reading is increasing from that when it is decreasing. Electrical and electronic instruments can suffer from drift of settings over a period of time owing to an increase in temperature of components.
2 **Operator skill**: a great many tests are based on personal manipulation of the apparatus and visual reading of a resultant indication. An op-

erator may be called on to prepare a sample, load it into the instrument, adjust readings such as zero and maximum, and to take a reading from a scale. Each manipulation, adjustment and reading involves an uncertainty which can depend on the skill and the conscientiousness of the individual operator. The ideal in instrument and test method design is to reduce the amount and scope of operator intervention.

3   **Fineness of scale division:** a fundamental limit is set to the precision of a measurement by the instrument scale which is necessarily subdivided at finite intervals. It carries with it an immediate implication of a minimum uncertainty of one half of the finest scale division. In the case of a digital scale the last digit of the display sets the limit to the precision in a similar manner as it has by its nature to be a whole digit. The final digit implies that it is plus or minus half of what would be the next digit. However, digital scales usually read to more figures than the equivalent analogue scale.

4   **External factors:** these may come from sources outside the actual instrument such as line voltage fluctuations, vibration of instrument supports, air currents, ambient temperature and humidity fluctuations and such diverse factors as variation in the sunlight intensity through windows.

The above uncertainties in the measurement of textile properties derive from the measurement process. In addition to these uncertainties, textile materials also exhibit variation in properties throughout their bulk. These can be quite considerable in magnitude, particularly in the case of yarns and fibres. This variability, in a similar manner to the errors described above, falls into two types: systematic, as is the case when the properties of a fabric vary from the edge to the centre, and random, when the variability has no pattern. The effect of this is to add to the errors from the measurement itself to give a larger overall error from which it is difficult to separate out the variability of the material from the experimental error. Therefore, because of all the above sources of variation, the appropriate statistical analysis of results has a great importance in textile testing.

1.4.6   Repeatability and reproducibility

The true accuracy of a test method can only be gauged by repeated testing of identical material both within the same laboratory and between different testing laboratories that possess the same type of equipment. International round trials [14–17] are organised by sending out sets of test samples, all produced from the same batches of material, to participating laboratories and asking them to test the samples in a prescribed manner. The results are then correlated and the within (repeatability) and between (reproducibility) laboratory variations calculated. The variation between laboratories is always greater than the variation found within a single laboratory.

BS 5532 [18] defines repeatability and reproducibility as follows.

Repeatability

1   *Qualitatively*: the closeness of agreement between successive results obtained with the same method on identical test material, under the same conditions (same operator, same apparatus, same laboratory and short intervals of time).
2   *Quantitatively*: the value below which the absolute difference between two single test results obtained in the above conditions may be expected to lie with a specified probability. In the absence of other indication, the probability is 95%.

Reproducibility

1   *Qualitatively*: the closeness of agreement between individual results obtained with the same method on identical test material but under different conditions (different operators, different apparatus, different laboratories and/or different times).
2   *Quantitatively*: the value below which the absolute difference between two single test results on identical material obtained by operators in different laboratories, using the standardised test method may be expected to lie with a specified probability. In the absence of other indication, the probability is 95%.

Errors involved

In order to understand the difference between repeatability and reproducibility the error in the test result can be considered to be due to two components [14]:

1   A random error (standard deviation σ_r) which occurs even when the same operator is using the same apparatus in the same laboratory. The variance of this σ_r^2 is called the within-laboratory variance and is assumed to have the same value for all laboratories.
2   An error (standard deviation σ_L) due to the difference that occurs when another operator carries out the test in a different laboratory using a different piece of identical apparatus. The variance of this σ_L^2 is called the between-laboratory variance.

The total error in a result that combines several sources of error can be obtained by adding together their variances. The numerical values for repeatability and the reproducibility are then given by substituting a value of 2 for n in the above equation for confidence limits:

$$\text{Repeatability} = \frac{2}{\sqrt{2}}\sigma_r$$

$$\text{Reproducibility} = \frac{2}{\sqrt{2}}(\sigma_L^2 + \sigma_r^2)^{1/2}$$

1.4.7 Significant figures

The numerical expression of the magnitude of a measurement may contain some figures that are doubtful. This can arise either from an estimation between the scale divisions by the operator or, in the case of a digital readout, from the uncertainty in the choice of the last figure by the machine. For instance in the case of a measurement of length by a rule that is graduated in millimetres, the rule might show that the length is definitely between 221 mm and 222 mm. Estimation by the person making the measurement might put the value at 221.6 mm. The figures 221 are exact but the final digit (6) is doubtful because it is only estimated. However, all four figures are regarded as significant because they convey meaningful information. This can be seen if it is imagined that the true value is actually 221.7 mm; the error would then be 0.1 mm but if the figures had been taken to the nearest whole millimetre (222) the error would have been 0.3 mm.

Significant figures, therefore, include all the exact figures followed by one doubtful one. Any zeros before the figures are not included in the number of significant figures and zeros after the figures are included only if they are considered to be exact regardless of the position of the decimal point. Zeros that are only there to position the decimal point are not regarded as significant; for example, 540,000 has only two significant figures. If it is necessary to express the fact that some of the zeros are significant, it is better to write the number as, for instance, 5.40×10^5. Zeros after a decimal point should be included only if they are significant. For instance the value 3.0 has two significant figures.

Unless otherwise indicated the uncertainty in any written measurement of a continuous variable is taken to be plus or minus half a step of the last significant figure. For instance 25.4 mm is taken to mean 25.4 ± 0.05 mm but 25.40 mm would be taken to mean 25.40 ± 0.005 mm.

The number of significant figures written down only concerns the reading of figures from instruments. It is an entirely separate issue from how many of the figures are meaningful which can only be decided from repeat tests as described above.

Rounding off

When further calculations are carried out on measured values the number of figures in a result may increase but in general the number that are significant does not increase. Retaining these figures in the final result gives a misleading impression of the precision of the result. The discarding of any figures beyond the significant ones is known as rounding off.

The convention for rounding off is that the last figure to remain is left unchanged if the amount to be discarded is less than 0.5, but it is increased by one if the amount to be discarded is greater than 0.5. For example: 6.854 would be rounded to 6.85 to three significant figures or 6.9 to two significant figures. Note that the rounding up or down is done only in one stage, not firstly to 6.85 and then to 6.8.

If the amount to be discarded is exactly 0.5 of a step then the rounding is to the nearest even figure in the last place, the idea being that this gives a random choice with as many results being rounded up as are rounded down. For example 6.85 would be rounded down to 6.8 whereas 6.95 would be rounded up to 7.0.

When carrying out calculations involving results with different numbers of significant figures, the number of figures in the result is governed by the contribution with the largest error. For example in addition or subtraction:

$$2.71 + 11.814 = 14.52$$
$$6.4 + 123.625 + 5.7165 = 135.7$$
$$2000 + 2,400,000 = 2,400,000$$

In each case the result is governed by the number whose concluding figure is the furthest to the left. In the case of simple multiplication or division the result should not in general be credited with more significant figures than appear in the term with the smallest number of significant figures. For example:

$$63.26 \times 0.0217 = 1.37$$
$$0.356 \times 0.6149 = 0.219$$

In case of doubt the mathematical operations can be carried out on the results for the implicit range of values.

Any rounding off must be carried out only on the final result after all the calculations have been made.

General reading

The WIRA Textile Data Book, WIRA, Leeds, 1973.
Morton W E and Hearle J W S, *Physical Properties of Textile Fibres*, 3rd edn., Textile Institute. Manchester. 1993.

Massey B S, *Measures in Science and Engineering*, Ellis Horwood, Chichester, 1986.

References

1. BS 2545 Methods of fibre sampling for testing.
2. IWTO-26-74 Glossary of terms relating to sampling.
3. BS 4784 Determination of commercial mass of consignments of textiles Part 2 Methods of obtaining laboratory samples.
4. ASTM D 2258 Sampling yarn for testing.
5. Lunney H W M, 'The pressure coring of wool bales', *Wool Sci Rev*, 1978 **55** 2.
6. ASTM D 1060 Core sampling of raw wool in packages for determination of percentage of clean wool fiber present.
7. Anon, 'The measurement of wool fibre length', *Wool Sci Rev*, 1952 **9** 15.
8. BS 2010 Method for determination of the linear density of yarns from packages.
9. ASTME, *Handbook of Industrial Metrology*, Prentice Hall, New Jersey, 1967.
10. Leaf, G A V, *Practical Statistics for the Textile Industry: Part 1*, Textile Institute, Manchester, 1984.
11. Leaf, G A V, *Practical statistics for the textile industry: Part 2*, Textile Institute, Manchester, 1987.
12. Murphy T, Norris K P and Tippett L H C, *Statistical Methods for Textile Technologists*, Textile Institute, Manchester, 1960.
13. ASTM D 2905 Statement on number of specimens for textiles.
14. Ly N G and Denby E F, 'A CSIRO Inter-laboratory trial of the KES-F for measuring fabric properties', *J Text Inst*, 1988 **79** 198.
15. Clulow E E, 'Comparison of methods of measuring wettability of fabrics and inter-laboratory trials on the method chosen', *J Text Inst*, 1969 **60** 14–28.
16. Lord J, 'Draft method for determination of dimensional changes during washing: interlaboratory trials', *Tex Ins Ind*, 1972 **10** 82.
17. Smith P J, 'Proposed new ISO domestic washing and drying procedures for textiles a background paper', *Tex Ins Ind*, 1977 **15** 329.
18. BS 5532 Part 1 Glossary of terms relating to probability and general terms relating to statistics.

2.1   Introduction

The properties of textile fibres are in many cases strongly affected by the atmospheric moisture content. Many fibres, particularly the natural ones, are hygroscopic in that they are able to absorb water vapour from a moist atmosphere and to give up water to a dry atmosphere. If sufficient time is allowed, equilibrium will be reached. The amount of moisture that such fibres contain strongly affects many of their most important physical properties. The consequence of this is that the moisture content of all textile products has to be taken into account when these properties are being measured. Furthermore because the percentage of moisture that can be retained by fibres is quite high (up to 40% with some fibres), the moisture content can have a significant effect on the mass of the material. This factor has a commercial importance in cases where material such as yarns and fibres is bought and sold by weight.

2.2   Effect of moisture on physical properties

The physical properties of fibres can be affected by their moisture content. In general the fibres that absorb the greatest amount of moisture are the ones whose properties change the most. Three main types of properties are affected.

2.2.1 Dimensional

The mass of the fibres is simply the sum of the mass of the dry fibre plus the mass of the water. The absorption of moisture by fibres causes them to swell, because of the insertion of water molecules between the previously tightly packed fibre molecules. Because the fibre molecules are long and narrow most of the available intermolecular spaces are along the length of the molecules rather than at the ends, so that the swelling takes place mainly in the fibre width as shown in Table 2.1. Nylon is a notable exception to this.

Table 2.1 The swelling of fibres due to moisture absorption [2]

Fibre	Transverse swelling		Longitudinal swelling (%)	Volume swelling (%)
	Diameter (%)	Area (%)		
Cotton	20, 23, 7	40, 42, 21		42, 44
Mercerised cotton	17	46, 24	0.1	
Viscose	25, 35, 52	50, 65, 67, 66, 113, 114	3.7, 4.8	109, 117, 115, 119, 123, 126, 74, 122, 127
Acetate	9, 11, 14, 0.6	6, 8	0.1, 0.3	
Wool	14.8–17	25, 26		36, 37, 41
Silk	16.5, 16.3–18.7	19	1.3, 1.6	30, 32
Nylon	1.9–2.6	1.6, 3.2	2.7–6.9	8.1–11.0

Table 2.1 is a summary of measurements made by different workers so that there is a certain amount of discrepancy among them. Because of the non-circular cross-section of a number of fibres, most notably cotton, the percentage change in cross-sectional area is a better measure than change in diameter. The change in volume of a fibre is linked to the changes in its length and cross-sectional area by simple geometry. The change in volume is also linked to the amount of water that has been absorbed. The swelling of fibres is a continuous process which takes place in step with their increasing moisture content. From this it follows that the swelling increases with the relative humidity of the atmosphere, the shape of the curve linking swelling to relative humidity being similar to that linking fibre regain with relative humidity [1].

Fabrics made from fibres that absorb large amounts of water are affected by the swelling. When such a fabric is soaked in water the increase in width of the fibres leads to an increase in diameter of the constituent yarns. Depending on the closeness of spacing of the yarns this can lead to a change in dimensions of the fabric. However, on subsequent drying out the structure does not necessarily revert to its original state. This behaviour is responsible for the dimensional stability problems of certain fabrics. Advantage is taken of fibre swelling in the construction of some types of waterproof fabrics whose structures are designed to close up when wetted, so making them more impermeable to water.

2.2.2 Mechanical

Some fibres, such as wool and viscose, lose strength when they absorb water and some, such as cotton, flax, hemp and jute, increase in strength.

Furthermore the extensibility, that is the extension at a given load, can increase for some fibres when they are wet. Figure 2.1 shows the loss in strength and the gain in elongation of a sample of wool tested when wet compared with a similar sample tested when dry. These changes in strength and extension have consequences for many other textile properties besides tensile strength. Some properties such as fabric tearing strength are ones that are obviously likely to be affected by fibre strength, but for other ones such as crease resistance or abrasion resistance the connection between them and changes in fibre tensile properties is less apparent. It is because of these changes in properties that textile tests should be carried out in a controlled atmosphere.

2.2.3 Electrical

The moisture content of fibres also has an important effect on their electrical properties. The main change is to their electrical resistance. The resistance decreases with increasing moisture content. For fibres that absorb water the following approximate relation between the electrical resistance and the moisture content holds for relative humidities between 30% and 90% [3, 4]:

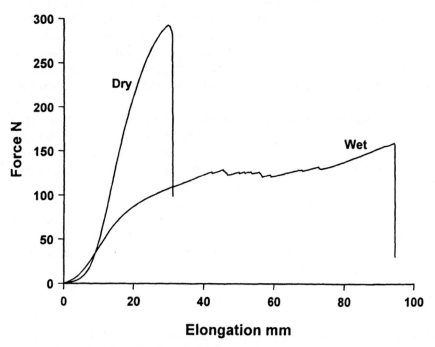

2.1   The strength of wet and dry wool.

$$RM^n = k$$

where R = resistance,

M = moisture content (%),

and n and k are constants.

The changes in resistance are large: there is approximately a tenfold decrease in resistance for every 13% increase in the relative humidity. Figure 2.2 shows the change in resistance with relative humidity for a sample of nylon [4]. This fall in resistance with increasing moisture content means that static electrical charges are more readily dissipated when the atmospheric relative humidity is high.

The relative permittivity (dielectric constant) of fibres increases with increasing moisture content in those fibres that absorb moisture [1]. Water itself has a much higher permittivity than the material making up the fibre and so as moisture is absorbed by the fibre the overall value is influenced by this, which will therefore affect any capacitance measurements, such as for evenness, which are made on textile materials.

2.3   Atmospheric moisture

The moisture content of textile materials when they are in equilibrium with their surroundings is determined by the amount of moisture in the air.

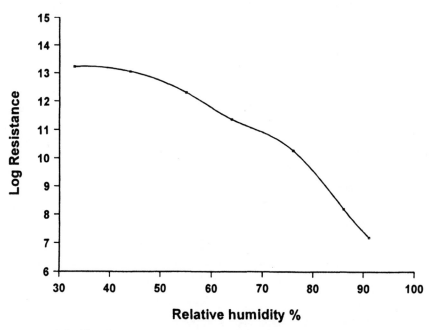

2.2   The change in resistance of nylon with relative humidity.

Therefore the moisture content of those materials that absorb water can vary from day to day or from room to room. The atmospheric moisture level is normally expressed in terms of relative humidity and not absolute water content.

2.3.1 Vapour pressure

Water molecules evaporate from the bulk at a rate determined by the exposed surface area and the temperature. Eventually the space above the surface reaches a stage when as many molecules are condensing back onto the surface as are evaporating from it. The space is then saturated with vapour. The amount of water held at saturation depends only on the temperature of the air and its value increases with increasing temperature as shown in Fig. 2.3. The pressure exerted by the water vapour is known as the saturated vapour pressure and is independent of the volume of space existing above the surface. If the vapour pressure in the space is kept higher than the saturated vapour pressure, water will condense back into the bulk. If, however, the vapour pressure is kept lower than the saturated vapour pressure, water will continue to evaporate from the surface until all the water has gone or the vapour pressures are equal. The total pressure above a surface is the pressure of the air plus the saturated vapour pres-

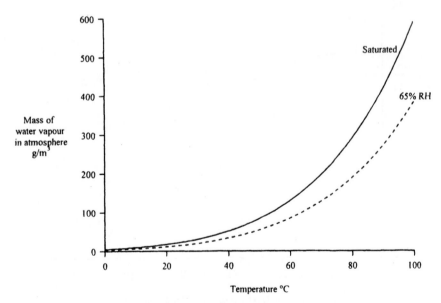

2.3   The mass of water vapour in the atmosphere (RH = relative humidity).

sure. For example at 20 °C the saturated vapour pressure of water is 17.5 mm Hg, if the atmospheric pressure is 760 mm Hg then the pressure of the air above a water surface is 742.5 mm Hg (Dalton's law of partial pressures).

2.3.2 Relative humidity

The amount of moisture that the atmosphere can hold increases with its temperature so that warmer air can hold more water than cold air. The converse of this is that when air containing moisture is cooled, a temperature is reached at which the air becomes saturated. At this point moisture will condense out from the atmosphere as a liquid: this temperature is known as the dew point.

When considering the effects of atmospheric moisture on textile materials the important quantity is not how much moisture the air already holds, but how much more it is capable of holding. This factor governs whether fibres will lose moisture to or gain moisture from the atmosphere. The capacity of the atmosphere to hold further moisture is calculated by taking the maximum possible atmospheric moisture content at a particular temperature and working out what percentage of it has already been taken up. This quantity is known as the relative humidity (RH) of the atmosphere and it can be defined in two ways. In terms of the mass of water vapour in the atmosphere:

$$RH = \frac{\text{mass of water vapour in given volume of air}}{\substack{\text{mass of water vapour required to saturate} \\ \text{this volume at the same temperature}}} \times 100\%$$

Alternatively it can also be defined as the ratio of the actual vapour pressure to the saturated vapour pressure at the same temperature expressed as a percentage:

$$RH = \frac{\text{actual vapour pressure} \times 100\%}{\text{saturated vapour pressure}}$$

The absolute humidity is defined as the weight of water present in unit volume of moist air measured in grams per cubic metre.

It is important to note that the relative humidity of the atmosphere changes with temperature even when the total quantity of water vapour contained in the air remains the same. The dotted line in Fig. 2.3 shows the increase in the mass of water vapour contained in the atmosphere with increasing temperature for a constant relative humidity of 65%.

The amount of moisture contained by fibres that are in equilibrium with the atmosphere is dependent on the relative rather than the absolute humidity.

2.3.3 Standard atmosphere

Because of the important changes that occur in textile properties as the moisture content changes, it is necessary to specify the atmospheric conditions in which any testing is carried out. Therefore a standard atmosphere has been agreed for testing purposes [5] and is defined as a relative humidity of 65% and a temperature of 20 °C. For practical purposes certain tolerances in these values are allowed so that the testing atmosphere is RH 65% ± 2%, 20 ± 2 °C. In tropical regions a temperature of 27 ± 2 °C may be used.

2.3.4 Measurement of atmospheric moisture

There are a number of different instruments for measuring the moisture content of the atmosphere, known as hygrometers or psychrometers.

Wet and dry bulb hygrometer

If the bulb of a glass thermometer is surrounded by a wet sleeve of muslin in an atmosphere that is not saturated, water vapour will evaporate into the air at a rate proportional to the difference between the actual humidity and 100% humidity. Owing to the latent heat of evaporation, heat is drawn from the thermometer bulb, thus cooling it. This cooling effect has the consequence that the temperature indicated by a wet bulb thermometer is lower than the air temperature. By mounting two identical thermometers together, one with a wet sleeve and one with a normal bulb, the two temperatures can be read directly. The relative humidity can then be calculated from the temperature difference between the two readings. The value is usually read from appropriate tables. The rate of evaporation of water is also governed by the speed of the airflow past the wet bulb. Therefore for accurate work the rate of airflow past the thermometer bulbs has to be controlled as still air conditions are difficult to achieve in practice. The sling and the Assmann type hygrometers are two instruments in which the flow of air is controlled.

Sling or whirling hygrometer

This instrument works on the same principle as above but the two thermometers are mounted on a frame with a handle at one end. This allows them to be rotated by hand at a speed of two or three revolutions per second, so giving an air speed of at least 5 m/s past the thermometers. After half a minute of rotation, temperature readings from the two thermometers are taken and the procedure is then repeated until the readings have reached minimum values.

Assmann hygrometer

This is a more sophisticated instrument than the sling hygrometer in that a fan is used to draw air across the thermometer bulbs at a constant pre-determined speed. The temperatures are read when they have reached a steady value.

Hair hygrometer

Human hair increases or decreases in length as the humidity of the sur-rounding air increases or decreases. By attaching a bundle of hairs to a suit-able lever system, the relative humidity of the atmosphere can be indicated directly and, if required, recorded on a chart. The accuracy of this method is limited to within 3 or 4% of the true value for the range of relative humidities between 30% and 80%. A combined temperature and humidity recording instrument is often used in laboratories and is known as a thermo-hygrograph. The hair hygrometer requires frequent calibration and has a slow response to changes in atmospheric conditions.

2.4   Regain and moisture content

The amount of moisture in a fibre sample can be expressed as either regain or moisture content. Regain is the weight of water in a material expressed as a percentage of the oven dry weight:

$$\text{Regain} = \frac{100 \times W}{D}\%$$

where D is the dry weight and W is the weight of absorbed water.

Moisture content is the weight of water expressed as a percentage of the total weight

$$\text{Moisture content} = \frac{100 \times W}{D + W}\%$$

Regain is the quantity usually used in the textile industry.

2.4.1   Regain – humidity relations of textiles

Hysteresis

If two identical samples of fibre, one wet and one dry, are placed in a stan-dard atmosphere of 65% RH, it might be expected that they would both eventually reach the same value of regain. However, this is not the case as the one that was originally wet is found to have a higher regain than the one that was originally dry; this is shown diagrammatically in Fig. 2.4. This difference is due to hysteresis between moisture uptake and moisture loss.

If the regain of a fibre that absorbs moisture is plotted against the atmospheric relative humidity as in Fig. 2.5 it is found to have an S – shaped curve and not a straight line relationship. If the relationship is plotted for decreasing relative humidity, that is when the fibres are drying out, it is found that the curve is different from that plotted for increasing relative humidity. It is this difference between the curves which is responsible for the difference in equilibrium regain values shown in Fig. 2.4. The dotted line in Fig. 2.5 at 65% relative humidity cuts the absorption and desorption curves at different values of regain. A sample that absorbed moisture in an atmosphere of 65% RH and reached equilibrium at the lower value of regain would then follow an intermediate path as it dried out. This phenomenon, which can result in two different values of regain at the same relative humidity depending on whether the sample is gaining or losing water, is important when samples are being conditioned. It is necessary for reproducibility of moisture content for a sample to approach equilibrium in the standard testing atmosphere from the same direction every time. Where this factor is important the samples are dried in an oven at a low temperature (50°C) before conditioning in the standard atmosphere.

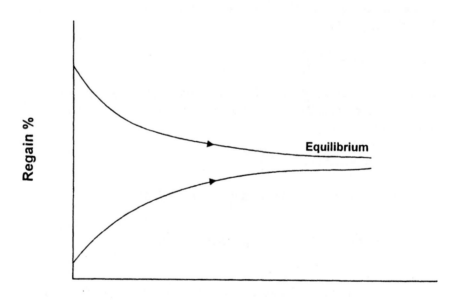

2.4 The hysteresis in moisture absorption.

2.4.2 Factors affecting the regain

Different fibre types absorb different amounts of moisture depending on their affinity for water as shown in Fig. 2.6 [6–10]. For a given fibre type the moisture content is governed by a number of factors:

1 **Relative humidity**. The higher the relative humidity of the atmosphere, the higher is the regain of textile material which is exposed to it.
2 **Time**. Material that is in equilibrium at a particular relative humidity which is then moved to an atmosphere with a different relative humidity takes a certain amount of time to reach a new equilibrium. The time taken depends on the physical form of the material and how easily the moisture can reach or escape from the individual fibres. For example the British Standard for count testing [11] suggests a period of one hour for yarn in hank form to reach equilibrium, but three hours for yarn on packages.
3 **Temperature**. For practical purposes the temperature does not affect the regain of a sample.
4 **Previous history**. The moisture content of textile materials in equilibrium with a particular relative humidity depends on the previous history of the material. For example the hysteresis effect as mentioned above

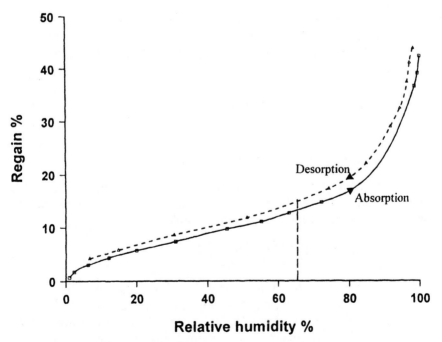

2.5 A plot of regain versus relative humidity for viscose fibres.

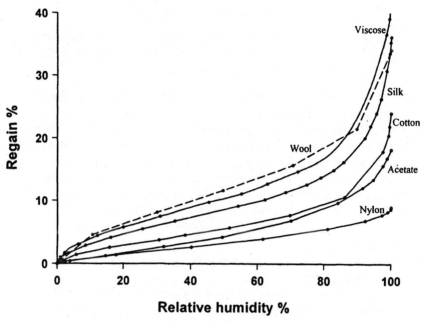

2.6 A comparison of the moisture uptake of fibres.

means that it will have a different moisture content depending on whether it was previously wet or dry. Processing of the material can also change its regain value by altering its ability to absorb moisture. The removal of oils, waxes and other impurities can also change the regain by removing a barrier on the fibre surface to the flow of moisture vapour. For example the standard regain value for scoured wool is 16% and that for oil combed tops is 19%.

2.4.3 Methods of measuring regain

To measure the regain of a sample of textile material it is necessary to weigh the material, dry it and then weigh it again. The difference between the masses is then the mass of water in the sample.

$$\text{Regain} = \frac{\text{mass of water} \times 100\%}{\text{oven dry mass}}$$

Regain is based on the oven dry mass, which for most fibres is the constant mass obtained by drying at a temperature of $105 \pm 2\,°C$. Constant mass is achieved by drying and weighing repeatedly until successive weighings differ by less than 0.05%. The relevant British Standard [12] specifies that successive weighings should be carried out at intervals of 15 min when using a ventilated oven, or at 5 min intervals if using a forced air oven. The exceptions to the above conditions are: acrylic fibres which should be dried in a normal oven at $110 \pm 2\,°C$ for 2 h and chlorofibres which should be dried at $77 \pm 2\,°C$ to constant mass.

Conditioning oven

A conditioning oven, as shown in Fig. 2.7, is a large oven which contains the fibre sample in a mesh container. The container is suspended inside the oven from one pan of a balance, the mechanism of which is outside the oven. This ensures that the weight of the sample can be monitored without the need to remove it from the oven. A continual flow of air at the correct relative humidity is passed through the oven which is maintained at $105\,°C$.

The main advantage of using a conditioning oven for carrying out regain determinations is that all the weighing is carried out inside the oven. This means that the sample does not gain moisture as it is taken from the oven to the balance. The oven is also capable of drying large samples. The use of a conditioning oven to dry a sample is the correct standard procedure; any other method of sample drying has to be checked for accuracy against it.

The method is based on the assumption that the air drawn into the oven is at the standard atmospheric conditions. If this is not the case then a correction has to be made based on the actual temperature and relative humidity of the surrounding air.

The basis of the recommended correction [13] is to add the following percentage to the dry weight of the sample

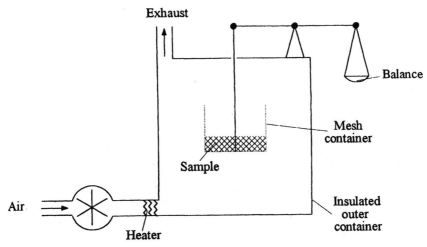

2.7 A conditioning oven.

$$\text{Percentage correction} = 0.5(1 - 6.48 \times 10^4 \times E \times R)\%$$

where R = relative humidity %/100,

E = saturation vapour pressure in pascals at the temperature of the air entering the oven (taken from a table of values).

Rapid regain dryer

The rapid regain type of dryer represents a quicker way of drying fibre samples. The basis of this type of dryer is that the hot air is blown directly through the sample to speed up the drying process.

There are a number of different versions of this instrument, but in all cases the sample being dried has to be removed from the output end of the hot air blower and weighed on a separate balance. In some instruments a removable sample container which has interchangeable end caps is used. One end cap is a perforated one for use when the air is being blown through and the other end caps are a pair of solid ones which are placed on the container when it is removed from the heater for weighing.

The WIRA improved rapid regain dryer, shown in Fig. 2.8, uses a powerful blower to force air through the sample in order to give more rapid drying. Because of the high volume of heated air which can be passed through the instrument a large sample can be dried. This can either be weighed in its container or removed from the container and weighed separately.

The CSIRO (Commonwealth Scientific and Industrial Research Organization) direct reading regain dryer is similar in principle but has a vertical column with a removable sample container at the top. There is also

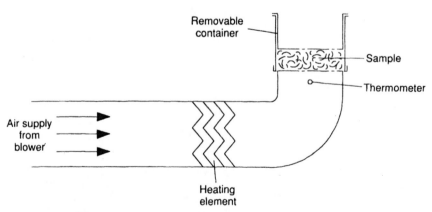

2.8 The WIRA process dryer.

a dedicated balance which forms part of the equipment and is calibrated to give the regain figure directly. Alternatively a balance may be connected to a computer which will calculate the regain from the weight. In order to achieve this the original sample is placed on the balance which is then zeroed before drying. As the sample then dries out, the balance indicates percentage regain.

When an external balance is used to weigh heated samples certain precautions have to be made:

1 The balance should be enclosed.
2 The time for weighing should not exceed 20s from removal of the sample from the dryer.
3 A buoyancy correction should be determined by using a dummy sample of steel wool which is weighed when cold and then reweighed after heating for a set time (5 or 15 min).

Electrical methods

The electrical properties of fibres change quite markedly with their moisture content so that the measurement of resistance or capacitance changes can be used to give an indirect method of regain determination.

The resistance change is a more suitable basis for an instrument as there is a greater change in the resistance of fibres with moisture content than there is in the capacitance. Furthermore the weight and distribution of material between the plates of a capacitor must be closely controlled in order to give reproducible results.

The great advantages that electrical methods possess over drying and weighing methods are the speed and ease of reading, the fact that they can be calibrated directly in regain units and the ease with which instruments can be made portable.

The disadvantages of electrical methods are the need to recalibrate them as they are indirect methods, the variations in readings due to packing density, the possible presence of dyes, antistatic agents and also variations in fibre quality.

In a typical electrical resistance measuring instrument, two electrodes are pushed into a package of yarn and the resistance between the electrodes is measured by suitable electronics, the answer being displayed on a scale which is directly calibrated in regain values. Different electrode sets are used for different packages, for example long thick prongs for bales and short needle like probes for yarn packages. The instrument usually has to be calibrated for the type of probe, the type of fibre and the expected regain range.

2.5   Correct invoice weight

When textile materials are bought and sold by weight, it is necessary for there to be agreement between buyer and seller on the exact weight that has to be paid for. This value can vary considerably with the moisture content of the material which in turn varies with type of material, the atmospheric moisture content at the time and how wet or dry the material was before it was packed, among other factors. The buyer certainly does not wish to pay for excess water at the same price per kilogram as the textile material. A 'correct invoice weight' is therefore determined according to [12]. In this procedure the consignment is considered to contain a percentage of water known as the standard regain allowance and the weight of the consignment is calculated as if it contained this amount of water.

When a consignment of textile material is delivered and weighed, a sample is taken from it on which tests are made which enable the correct invoice weight to be calculated. Samples of at least 200 g are selected according to adequate sampling procedures and immediately stored in airtight containers so that no moisture is lost. The samples are weighed and then the oven dry weight is determined as described above. In some cases other non-textile materials, such as oils, grease, wax and size, are removed before drying.

If M = mass of consignment at time of sampling, D = oven dry mass of sample, S = original mass of sample and C = oven dry mass of the consignment:

$$C = M \times \frac{D}{S}$$

To the oven dry mass is added an official allowance for moisture depending on the nature of the material. This regain allowance, sometimes called the 'official' or 'standard' regain, is set out in BS 4784. These values are only approximately the regains the materials would have when in equilibrium with the standard atmosphere and represent agreed commercial numbers for the purposes of determining quantities such as consignment weights, yarn counts and percentage compositions which vary with moisture content.

The regain allowances vary depending on what physical state the material is in, for example woollen yarn 17%, worsted yarn 18.25%, oil combed tops 19%, wool cloth 16%.

$$\text{Correct invoice weight} = C \times \left(\frac{100 + R_1}{100} \right)$$

where R_1 = commercial moisture regain.

Table 2.2 Regain allowances. Note that regain values depend on the form of the material

Fibre type	UK regain (%)	US regain (%)
Man-made fibres		
Acetate	(9)	6.5
Acrylic	2	1.5
Nylon 6, 6 and 6	(6.25)	4.5
Polyester	(1.5)	0.4
Polypropylene	(2)	0
Triacetate	(7)	3.5
Viscose	(13)	11
Natural fibres		
Cotton – natural yarn	8.5	7
Linen fibre	12	12
Linen yarn	12	8.75
Silk	11	11
Wool – worsted yarn	18.25	13.6
Wool – fibre clean scoured	17	13.6

Figures in brackets commercial allowances for cleaned fibres.
Source: taken from [12] and [13].

If the samples are dried after cleaning a different set of allowances is used for moisture and oil content, etc:

$$\text{Correct invoice weight} = C \times \left(\frac{100 + R_2 + A_2 + B_2}{100} \right)$$

where R_2 is the moisture regain which may differ from R_1, A_2 is the allowance for natural grease and B_2 is the allowance for added oil. In most cases an overall allowance is given which includes the values for moisture and natural and added fatty matter.

In the case of a blend the overall allowance is calculated from the fraction of each component in the blend multiplied by its regain value, for example: 50/50 wool / viscose (dry percentages)

$$\frac{50 \times 17}{100} + \frac{50 \times 13}{100} = 8.5 + 6.5 = 15\%$$

Regain values vary slightly from country to country. For instance, the USA has a single value of 13.6% for wool in all its forms but has separate values for natural cotton yarn (7.0%), dyed cotton yarn (8.0%) and mercerised cotton yarn (8.5%), although there is no value laid down for raw cotton. Therefore the appropriate standard should be consulted for the correct commercial regain figures. Table 2.2 is intended only as a guide.

Table 2.3 The relative humidity of air over saturated solutions of salts at 20 °C

Saturated salt solution	Relative humidity (%)
Potassium sulphate	97
Potassium nitrate	93
Potassium chloride	86
Ammonium sulphate	81
Sodium chloride	76
Sodium nitrite	66
Ammonium nitrate	65
Sodium dichromate	55
Magnesium nitrate	55
Potassium carbonate	44
Magnesium chloride	33
Potassium acetate	22
Lithium chloride	12
Potassium hydroxide	9

2.6   Control of testing room atmosphere

Testing laboratories require the atmosphere to be maintained at 65 ± 2% RH and 20 ± 2 °C in order to carry out accurate physical testing of textiles. The temperature is controlled in the usual way with a heater and thermostat, but refrigeration is necessary to lower the temperature when the external temperature is higher than 20 °C as is usually the case in summer. The tolerances allowed on temperature variation are quite difficult to meet. The relative humidity is controlled by a hygrometer which operates either a humidification or a drying plant depending on whether the humidity is above or below the required level. Double glazing and air locks to the doors are usually fitted to the laboratory in order to reduce losses and to help to keep the atmosphere within the tolerance bands. Adequate insulation is necessary on all external surfaces as the high moisture content of the atmosphere can cause serious condensation on cold surfaces. Only large organisations may be able to afford a fully conditioned testing laboratory. If one is not available, testing should be carried out in a room in which ambient conditions are as uniform as possible throughout the year in order to cut down on variations in measurements due to atmospheric variation.

Cabinets that control the atmosphere of a relatively small volume can be obtained commercially. These may be used to condition samples before testing, the actual tests being carried out in a normal atmosphere straight after removal from the conditioned atmosphere. The relative humidity of a small enclosed volume of air, such as a desiccator, may be controlled by the presence of a dish containing a saturated solution of certain salts such as those listed in Table 2.3 [14].

References

1. Morton W E and Hearle J W S, '*Physical Properties of Textile Fibres*', 3rd edn, Textile Institute, Manchester, 1993.
2. Preston J M and Nimkar M V, 'Measuring the swelling of fibres in water', *J Text Inst*, 1949 **40** P674.
3. Hearle J W S, 'The electrical resistance of textile materials: I The influence of moisture content', *J Text Inst*, 1953 **44** T117.
4. Cusick G E and Hearle J W S, 'The electrical resistance of synthetic and cellulose acetate fibres', *J Text Inst*, 1955 **46** T699.
5. BS EN 20139 Textiles. Standard atmospheres for conditioning and testing.
6. Urquhart A R and Eckersall N, 'The absorption of water by rayon', *J Text Inst*, 1932 **23** T163.
7. Urquhart A R and Eckersall N, 'The moisture relations of cotton VII a study of hysteresis', *J Text Inst*, 1930 **21** T499.
8. Speakman J B and Cooper C A, 'The adsorption of water by wool I Adsorption hysteresis', *J Text Inst*, 1936 **27** T183.
9. Hutton E A and Gartside J, 'The moisture regain of silk I Adsorption and desorption of water by silk at 25 °C', *J Text Inst*, 1949 **40** T161.
10. Hutton E A and Gartside J, 'The adsorption and desorption of water by nylon at 25 °C', *J Text Inst*, 1949 **40** T170.
11. BS 2010 Method for determination of the linear density of yarns from packages.
12. BS 4784 Determination of commercial mass of consignments of textiles Part 1 Mass determination and calculations.
13. ASTM D 1909 Table of commercial moisture regains for textile fibres.
14. The *WIRA Textile Data Book*, WIRA, Leeds, 1973.

3
Fibre dimensions

3.1 Fibre fineness

Fineness is one of the most important properties of the fibres that are made into textile products. The fibre fineness has a number of effects on the properties of the yarn and hence the fabric that is made from it. The finer the fibre, the finer is the yarn that can be spun from it. As the yarn becomes thinner, the number of fibres in its cross-section decreases and the yarn becomes increasingly uneven because the presence or absence of a single fibre has a greater effect on the yarn diameter. The spinning limit, that is the point at which the fibres can no longer be twisted into a yarn, is reached earlier with a coarser fibre. Alternatively for a yarn of a given linear density the use of a finer fibre will enable the production of a more even yarn.

When staple fibres are twisted together into a yarn the twist provides the force that holds the individual fibres together. Less twist is needed to make a yarn of a given strength from fine fibres as the greater surface area which they possess provides more cohesion.

The most important effect of fibre fineness is on the fibre stiffness. This is because the rigidity of a fibre increases with the fourth power [1] of the fibre diameter so that a coarser fibre is a great deal stiffer. The stiffness of the fibres affects the stiffness of the fabric made from it and hence the way it drapes and how soft it feels. It also affects how soft or how prickly the fabric feels when it is worn next to the skin. This is because the skin touches the fibres that stick up from the surface of a fabric. If they are stiff they require more force to bend them and the skin feels this as prickle. Contrast the stiffness of a nylon monofilament when it is used for the bristles of a brush with the same material but with a much finer diameter when it is used in tights or stockings.

Fineness of fibres is a highly prized commodity which enables garments to be made with a soft and luxurious handle. With natural materials such as cotton, silk, wool and other animal fibres the finer varieties are reserved for the more expensive apparel and hence command higher prices. This is the main reason that fibres such as cashmere and silk which are naturally fine are comparatively expensive. It is therefore important commercially to be able to determine the fineness of natural fibres as this is an important factor in their quality and hence price. With natural materials the fibre diameter does not have a single value but it has a fairly wide distribution of sizes even in material of one type, for example wool from a single sheep.

3.2   Fineness measurement

When considering ways of measuring fibre fineness there are a number of factors that need to be taken into account which make it difficult to define a measure of fineness that is applicable to all fibres:

1   The cross-section of many types of fibres is not circular. Wool has an approximately circular cross-section but silk has a triangular cross-section and cotton is like a flattened tube as shown in Fig. 3.1. Man-made fibres are often made with trilobal (Fig. 3.2), star or hollow cross-sections for particular purposes. This makes it impossible to have a universal system of fibre fineness based on fibre diameter.
2   The cross-sections of the fibres may not be uniform along the fibre length. This is often the case with natural fibres.

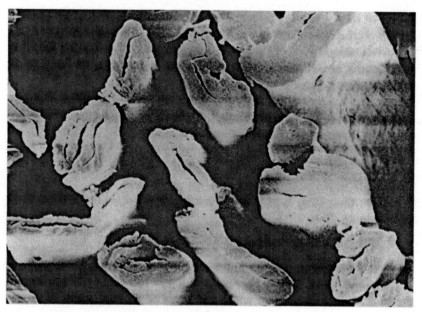

3.1   Cross-section of cotton fibres ×1500.

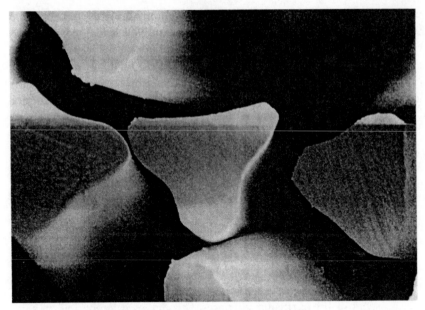

3.2   Cross-section of trilobal nylon fibres ×1000.

3   The cross-sectional shape of the fibres may not be uniform from fibre to fibre.

Because of these problems a definition of fibre fineness is needed that can allow for all the variations but that leads to a method of measurement which is relatively simple to carry out. The great degree of variability found in natural fibres means that a large number of measurements have to be carried out in such cases.

There are a number of different ways of measuring fibre fineness/diameter which differ fundamentally in their definitions of fineness so that the measurements are not easily interconvertible.

3.2.1 Gravimetric

For a given fibre (that is of a fixed density) its mass is proportional to its cross-sectional area:

Mass of a fibre = cross-sectional area × length × density

Therefore for a known length of fibre its mass will be directly related to its cross-sectional area. This relationship is made use of in the gravimetric definition of fibre fineness in which the mass of a given length of fibre is

Table 3.1 Fibre densities [2]

Fibre	Density g/cm³
Polypropylene	0.90
Polyethylene	0.95
Nylon 11	1.10
Nylon 6	1.13
Nylon 6,6	1.14
Acrylic	1.14–1.18
Triacetate	1.30
Wool	1.31
Acetate	1.33
Silk	1.33
Cotton	1.35
Polyester	1.38
Viscose	1.52

used as a measure of its fineness. This is similar to the system of measuring yarn linear density. The primary unit is tex (g/1000 m), but it is also common to use:

Decitex = mass in grams of 10,000 metres of fibre
Millitex = mass in milligrams of 1000 metres of fibre
Denier = mass in grams of 9000 metres of fibre

For fibres with a circular cross-section such as wool the mass per length can be converted into an equivalent fibre diameter sometimes known as d(grav.) using the following equation:

$$\text{Decitex} = 10^6 = \rho \times A$$

where A = cross-sectional area in cm²
       ρ = density in g/cm³
See Table 3.1 for a list of common fibre densities.
For a circular fibre then:

$$\text{Decitex} = 10^{-2} \times \rho \times \frac{\pi d^2}{4}$$

where d = diameter of fibre in micrometres
This reduces to:

$$\text{Decitex} = 7.85 \times 10^{-3} \times \rho \times d^2$$

Similarly for denier:

$$\text{Denier} = 7.07 \times 10^{-3} \times \rho \times d^2$$

For fibres with cross-sectional shapes other than circular the relationship is more complex as it is necessary to calculate the fibre cross-sectional area A.

When fibres have a cross-section that is ribbon-like, that is with one dimension smaller than the other, the fibre stiffness is determined by the smaller dimension as the fibre tends to bend in this direction when stressed. Fibres that have indented shapes such as star or trilobal cross-sections in which the maximum diameter is roughly the same in all directions are nearly as stiff as a fibre whose diameter is the same as their greatest dimension but have a lower mass per unit length and hence decitex value owing to the indentations. The strength of a fibre, however, is determined by the amount of material in the cross-section and not how it is arranged. These factors mean that when comparing fibres of different cross-section their decitex values are not necessarily a guide to their properties. Similarly fibres of different densities but of the same decitex value will have different diameters (assuming that their cross-sectional shape is the same).

In order to measure the mass per unit length of fibres a large number of them have to be cut accurately into short lengths and weighed on a sensitive balance. One way of doing this is to use two razor blades set into a holder so that they are accurately spaced 10 mm apart.

3.2.2 Fibre fineness by projection microscope

The projection microscope is the standard method [3] for measuring wool fibre diameter, and all other methods have to be checked for accuracy against it. The method is also applicable to any other fibres with a circular cross-section. The method involves preparing a microscope slide of short lengths of fibre which is then viewed using a microscope that projects an image of the fibres onto a horizontal screen for ease of measurement. The apparatus is shown diagrammatically in Fig. 3.3. Techniques are followed that avoid bias and ensure a truly random sample.

Method of test

A suitable random and representative sample is conditioned for 24 h in a standard testing atmosphere. Using a modified Hardy microtome the fibres are cut to a suitable length (0.4 mm for fibres below 27 μm) and a slide is prepared by carefully mixing the fibres into the mountant. The use of short fibres gives a length-biased sample so that proportionally more of the longer fibres will have their diameter measured. The mounting agent should be non-swelling and have a suitable refractive index (for example liquid paraffin). The mixture of fibres and mountant is spread thinly on the slide

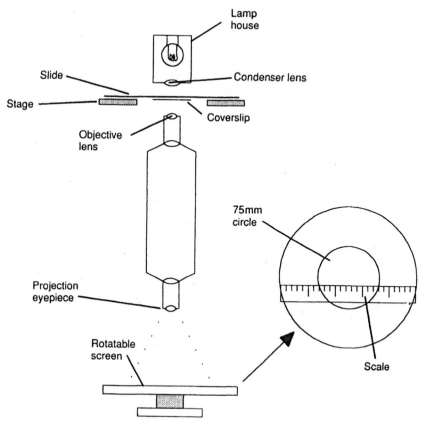

3.3   The projection microscope.

and covered with a cover glass, carefully avoiding air bubbles and finger prints.

The slide is placed on the stage, coverglass down (microscope inverted) and fibres are selected for measurement in the following way. The slide is traversed in a zigzag fashion as shown in Fig. 3.4, measuring every fibre that complies with the following requirements:

1   has more than half its length visible in the 7.5 cm circle which is drawn in the centre of the field of view;
2   is not in contact with any other fibre at the point of measurement.

The traverse of the slide is continued until the required number of fibres has been measured.

The magnification of the microscope is adjusted to be 500× so that on the scale used to measure the fibres each millimetre represents 2 μm.

Table 3.2 Number of measurements required for
a given accuracy

Confidence limit	Number of measurements
1%	2500
2%	600
3%	300
5%	100

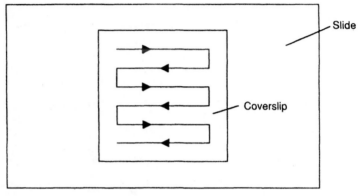

3.4   The path taken when viewing a projection microscope slide.

For accurate tests three slides should be measured from randomly selected areas of the material and not less than 150 fibres per slide should be measured.

The coefficient of variation of diameter for unblended wool lies between 20% and 28%. From this value the number of tests to give certain confidence limits has been calculated and is shown in Table 3.2.

3.2.3 Fibre fineness by the airflow method

This is an indirect method of measuring fibre fineness which is based on the fact that the airflow at a given pressure difference through a uniformly distributed mass of fibres is determined by the total surface area of the fibres [4]. The surface area of a fibre (length × circumference) is proportional to its diameter but for a given weight of sample the number of fibres increases with the fibre fineness so that the specific surface area (area per unit weight) is inversely proportional to fibre diameter; Fig. 3.5 shows this diagrammatically. Because the airflow varies with pressure difference it is the ratio of airflow to differential pressure that is determined by the fibre diameter. Therefore the method can be used to measure either

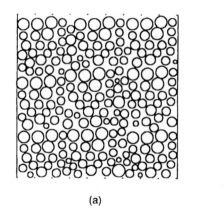
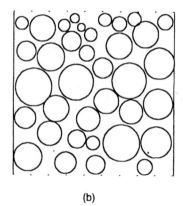

(a)   (b)

3.5   Airflow through coarse and fine fibres.

the airflow at constant pressure or the pressure drop at constant airflow. The measurement of airflow at constant pressure is the more usual form of apparatus with wool.

For fibres of approximately circular cross-section and constant overall density such as unmedullated wool, the estimate of fineness corresponds to the average fibre diameter as determined by the projection microscope with a good degree of accuracy.

Method of test for clean wool sliver

A random and representative sample of fibre is taken from the bulk and conditioned for 24 h in a standard testing atmosphere.

Three 2.5 g samples (2.5 ± 0.004 g) are prepared by cutting the sliver obliquely. Each sample in turn is evenly packed into the cylinder of the measuring instrument shown in Fig. 3.6. The cylinder is of a fixed size with a perforated base to allow the air to pass through. The wool is packed into the cylinder using a rod of a specified length so that the wool is not compressed to a greater density than specified. The perforated retaining cap is then screwed into position.

A pump is used to suck air through the sample and the pressure drop across the sample is monitored with a manometer. The manometer has a lower engraved mark to which the pressure is adjusted by a valve so that the airflow is measured at a fixed pressure drop. The volume of air flowing through the sample is then measured with a flowmeter.

Each sample is removed and retested four times and a mean is taken of the 12 readings.

The airflow is measured in arbitrary units which are then converted to micrometres using tables supplied with the instrument. These tables are specific for different types of wool such as oil or dry combed tops because

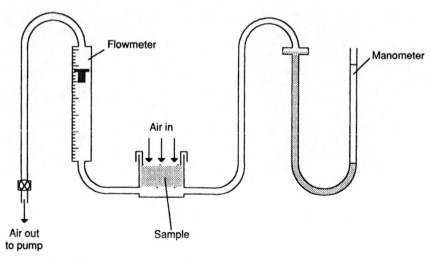

3.6   Fibre diameter measurement by airflow.

the surface condition of the fibres affects the airflow over them.

The machine is calibrated with a set of reference wools which have been measured by the projection microscope method.

Airflow measurements on core samples of raw wool

The above method was designed for clean wool top and is very specific for the condition of the fibres because of the effects this has on the ease of airflow through the fibre mass. However, the method has been adapted for core samples of raw wool [5] which require a cleaning process prior to testing. The machine also requires calibrating specially for core samples.

When core samples are made from a bale of raw wool the length of the fibres is reduced by the coring action. This has the effect of giving higher than expected airflow readings on an instrument calibrated for combed sliver. The samples are therefore tested using an airflow apparatus that has been calibrated using fibres chopped into 20 mm lengths.

Before test the cored samples undergo a fixed scouring process using water and a detergent and are oven dried. They are then carded using a Shirley analyser, dried and conditioned.

3.2.4   Cotton fineness by airflow

The method used for measurement of cotton fibre fineness by airflow [6, 7] is similar to that used for wool. However, the measurement is complicated by the fact that the results are affected by the maturity of the cotton fibres as well as by their fineness. Because of this the test results for the simple form of the instrument are usually expressed in arbitrary Micronaire units. (Micronaire is the trade mark of the Sheffield Corporation.)

The test is most frequently carried out on raw cotton which has to be opened and blended using a laboratory blender or Shirley analyser.

The mass of the sample must be accurately determined for the particular instrument being used. At least two samples are used with each sample being tested at least twice. The measurements are given in arbitrary Micronaire units which can be converted to the product of linear density (millitex) and maturity ratio.

3.2.5 Cotton maturity

Cotton fibre maturity denotes the degree of wall thickening of the fibres [8]. When the cotton fibres are first formed they start as thin tubules which initially grow only in length. When their maximum length is reached, a secondary fibre wall then begins to build up on the inner surface of the thin primary wall. This process continues until shortly before the boll opens. After the opening the fibres dry and collapse to give the typical convoluted ribbon form of cotton.

It was originally considered that the fibres with thinner walls had not matured for the same length of time as the normal fibres, hence the term maturity, but it is now known that the incomplete wall thickening is due to poor growth conditions. However, all cotton samples, even those grown under optimal conditions, contain a percentage of immature fibres.

Because immature fibres have thinner walls, their physical properties are different, they are weaker and less stiff than the mature fibres. This can lead to faults in articles made from cottons containing a high percentage of immature fibres. The faults of the immature fibres include: breaking during processing, a tendency to form neps, a tendency to become entangled around particles of trash and leaf, all adversely affecting yarn and fabric appearance. They may also dye to a lighter shade than the mature fibres.

For research purposes wall thickness can be denoted by degree of thickening θ

$$\theta = \frac{\text{cross-sectional area of fibre wall}}{\text{area of circle of same perimeter}}$$

A completely solid fibre would have a degree of thickening of 1. Mature fibres have an average value of around 0.6 and immature fibres have an average value of between 0.2 and 0.3.

To measure the maturity a sample of cotton is swelled in 18% sodium hydroxide and then examined under the microscope. The appearance of the

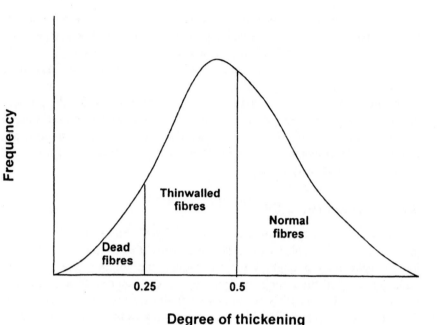

Degree of thickening

3.7   The distribution of fibre wall thickness in cotton.

swollen fibres depends on its degree of thickening and the fibres are classified into three groups according to their visual appearance:

1 *Normal fibres* are those that after swelling appear as solid rods and show no continuous lumen.
2 *Dead fibres* are those that after swelling have a continuous lumen and the wall thickness is a fifth or less than the ribbon width.
3 *Thin-walled fibres* are those that are not classed as normal or dead, being of intermediate appearance and thickening.

The distribution of the different types is shown in Fig. 3.7.
The results are expressed as the average percentages of normal (N) and dead (D) fibres from which the maturity ratio (M) is calculated:

$$M = \frac{(N - D)}{200} + 0.7$$

Maturity ratio is the measurement that is used commercially. It is connected to average degree of thickening by the relation:

Average degree of thickening = 0.577 M

In the USA the way of measuring maturity is different [8]; the fibres are assigned to just two classes:

1 *Mature fibres* – these in the swollen state have a ratio of apparent wall thickness to ribbon width that is greater than a quarter.

2 *Immature fibres* – the rest.

The mature class in the US test covers all the normal and part of the thin-walled group of the UK test. The result is expressed as percentage mature fibres:

$$P_M = \frac{\text{no. of mature fibres}}{\text{total no. of fibres}} \times 100\%$$

This value can be converted to maturity ratio by the following empirical relation:

$$M = 1.76 - \sqrt{(2.44 - 0.0212P_M)}$$

3.2.6 IIC / Shirley fineness and maturity tester

The simple Micronaire value of a cotton sample is dependent on both the fibre fineness and its maturity. Both higher maturity and coarser fibres can give a high Micronaire reading and conversely both fine fibres and immature fibres can give a low Micronaire reading. Therefore a particular reading could arise from a variable combination of the two factors. In practice the maturity of the cotton has a greater effect on its Micronaire value than its fineness. A particular cotton variety usually varies within ± 5% in fineness but its maturity may vary from about 10% above average to 20% less than average. The variation in maturity has a magnified effect on the Micronaire value as this is dependent on the maturity ratio squared; consequently the Micronaire value can vary from 20% above to 40% below its typical value. Therefore for a given cotton variety the Micronaire value is mainly a measure of its maturity. This, however, does not apply when different varieties of cotton are being compared.

In view of the above interdependence the fineness – maturity tester was developed by the Shirley Institute for the International Institute for Cotton. The test is based on determining the airflow through a cotton sample under two different sets of conditions which enables the equations to be solved for both fibre fineness and maturity. In the equipment the air permeability is measured at two different compression densities of the sample at particular pre-set rates of airflow for each. The air pressure is measured in both cases as this is more accurate than keeping the pressure constant and measuring the airflow. The pre-set rate of flow at the initial sample density is higher than that at the second, higher sample density. These measurements enable both the fineness and maturity to be calculated from the same sample. The specimen density used at the lower, initial compression is similar to that used in the Micronaire test.

3.2.7 Optical fibre diameter analyser

The determination of the fibre diameter of wool by manual operation of a projection microscope is a far from ideal test method. The procedure is slow, tedious and subject to operator variability. However, it provides data on fibre diameter distribution whereas the quicker and easier airflow test only measures the mean diameter. Not surprisingly, many attempts have been made over the years to devise an improved method of determining the diameter distribution of fibres in order to speed up the process.

The OFDA (optical fibre diameter analyser) [9] is a microscope-based system which effectively automates the projection microscope. The microscope uses a stage that is driven by two stepper motors under computer control to give an X–Y scan of the slide. The image is collected with a video camera using a relatively low powered objective in order to minimise focusing problems. The image is digitised by a frame grabber board in the computer to give a real time image with a 256 × 256 pixel matrix. Pattern recognition software identifies and measures fibres to a resolution of 1 μm and puts them into 1 μm groups. The whole image from an area of about 1 mm × 1.5 mm is analysed at a time; this may contain between 3 and 50 fibres. Sixteen images can be analysed per second so that a whole measurement on a sample can be completed in less than a minute. Preparation of the samples is vital for the success of the method. Fibres are cut into 1.5–2.0 mm snippets and spread using a purpose-built spreader onto 70 mm × 70 mm special glass slides. These are assembled in pairs with a fabric hinge.

3.2.8 Light-scattering methods

The fibre diameter analyser (FDA) [10, 11] system is a non-microscopical method of measuring fibre diameter which operates by light scattering. In the instrument the fibres are caused to intersect a circular beam of light in a plane at right angles to the direction of the beam. As the fibre passes through the beam the intensity of the scattered light reaches a maximum which is closely proportional to the projected area of the fibre. Only fibres that completely cross the beam are recorded so that the scattered light pulse is then proportional to the fibre diameter. The beam diameter is no greater than 200 μm in order to reduce the effect of curvature of the fibres due to crimp.

In order to present the fibres to the beam in the correct manner they are cut into short snippets 1.8 mm long and suspended in isopropanol to give a slurry. This is circulated through a square section channel 2 mm deep, as shown in Fig. 3.8, at a suitable flow rate and concentration so that they intersect the beam one at a time. A proportion of the snippets do not fully

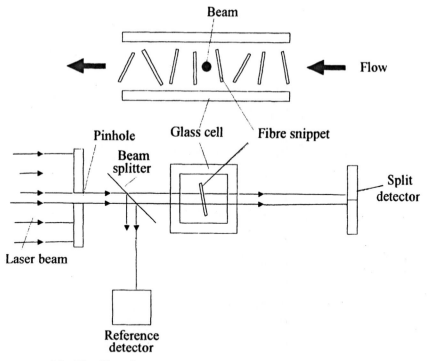

3.8 The fibre diameter analyser.

intersect the beam, and these are rejected by using a detector which is split into two halves, each operating on one half of the beam. If a fibre does not fully extend across the beam the signals from the two detectors are unequal and so the result can be rejected. The system is capable of measuring 50 fibres per second and can produce a mean fibre diameter and a diameter distribution.

3.2.9 Vibration method

The natural fundamental frequency (f) of vibration of a stretched fibre is related both to its linear density and to the tension used to keep it tight:

$$f = \frac{1}{2l}\sqrt{\left(\frac{T}{M}\right)}$$

or

$$M = T\left(\frac{1}{2lf}\right)^2$$

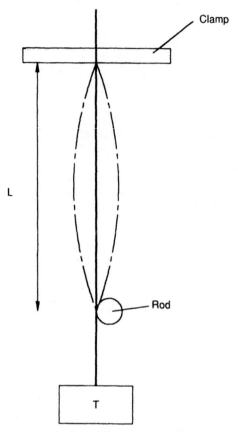

3.9   Fibre fineness measurement by vibration of a fibre.

where: M = mass per unit length,
     l = length,
     T = tension.

Vibroscopes have been constructed to measure the linear density of fibres by using this relationship [12]. In the basic form of the method as shown in Fig. 3.9, one end of a weighted fibre is clamped and the lower edge of the fibre passes over a knife edge, thus providing a fixed length of fibre under tension. The level of tension used is in the range from 0.3 to 0.5 cN/tex usually applied by hanging a weighted clip on the end of the fibre. The fibre is caused to vibrate either by vibrating the top clamp or by using acoustic transducers and the amplitude of the vibration measured over a range of frequencies. The frequency that gives rise to the maximum vibration amplitude is the fibre resonance frequency from which the linear density can be calculated. An alternative method for fibres with a narrow resonance peak is to excite the fibres close to this peak and then to discontinue the excitation, leaving the fibre to vibrate on its own. The free vibration frequency of the fibre corresponds to its resonance frequency.

3.2.10 Wool quality

The original basis for wool quality numbering was the grade given to it by an experienced wool sorter. The number represented his estimate of the finest worsted count yarn that could be spun from the sample of wool. For example a wool that was classed as 50s implied that a spinner could make a 50s count worsted yarn from it. Developments in spinning techniques have now enabled the spinner to produce finer counts than the quality number would suggest, so that the link between fibre quality and yarn count is no longer straightforward.

At one time the standards of wool quality were peculiar to a given mill as they were based on subjective criteria, but as ways of measuring the fibre diameter were developed, a number of scales relating quality to diameter were published. Because of this multiplicity of scales the International Wool Textile Organisation urged their discontinuation and the use of micrometer measurements alone for measuring wool fineness. Therefore no official international wool quality scale has ever existed and the increasing use of objective measurements for all the various aspects of quality has removed the need for one. However, the quality numbers are still used as descriptive terms in the wool industry so that the figures in Table 3.3 [2] are given as a guide.

Wool quality is often roughly divided into three main areas:

> Merino 60s and above
> Crossbred 36s to 58s
> Carpet up to 36s

3.3   Fibre length

After fineness, length is the most important property of a fibre. In general a longer average fibre length is to be preferred because it confers a number of advantages. Firstly, longer fibres are easier to process. Secondly, more even yarns can be produced from them because there are less fibre ends in a given length of yarn. Thirdly, a higher strength yarn can be produced from them for the same level of twist. Alternatively a yarn of the same strength can be produced but with a lower level of twist, thus giving a softer yarn.

The length of natural fibres, like their fineness, is not constant but it has a range of values even in samples taken from the same breed of animal or

Table 3.3 Wool grades

Grade	Average fibre diameter range (μm)
Finer than 80s	Under 18.10
80s	18.10–19.59
70s	19.60–21.09
64s	21.10–22.59
62s	22.60–24.09
60s	24.10–25.59
58s	25.60–27.09
56s	27.10–28.59
54s	28.60–30.09
50s	30.10–31.79
48s	31.80–33.49
46s	33.50–35.19
44s	35.20–37.09
40s	37.10–38.99
36s	39.00–41.29
Coarser than 36s	Over 41.29

plant. Man-made fibres on the other hand can be cut during production to whatever length is required with either all the fibres having the same length or with a distribution of lengths.

Cotton is a comparatively short fibre with the finest variety, Sea Island cotton, only reaching just over 50 mm (2 in) in length, whereas some varieties of Indian cotton may be less than 12 mm ($\frac{1}{2}$ in) long.

Wool is a much longer fibre than cotton and its length varies with the breed of sheep. The length can vary in from about 375 mm (15 in) long 36s quality in the Lincoln breed to 137–150 mm long ($5\frac{1}{2}$–6 in) in 64s quality Merino, the finer 80s and 90s qualities are 87–112 mm long ($3\frac{1}{2}$–$4\frac{1}{2}$ in).

As a general rule the longer wools are coarser than the shorter ones whereas in the case of cotton the longer fibres are finer than the short ones.

3.3.1 Mean length

In the case of natural fibres the definition of mean length is not as straightforward as it might at first seem. This is because natural fibres besides varying in length also vary in diameter at the same time. If the fibres all had the same cross-section then there would be no difficulty in calculating the mean fibre length. All that would be necessary would be to add up all the individual fibre lengths and divide this sum by the number of fibres. However, if the fibres have different diameters then the thicker fibres will have a greater mass so that there is a case for taking the mass into account

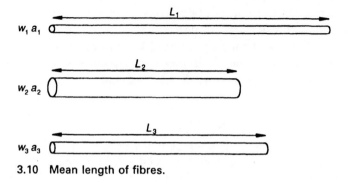

3.10   Mean length of fibres.

when calculating the mean length. There are in fact three possible ways of deriving the mean length:

1 Mean length based on number of fibres (unbiased mean length) L.
2 Mean length based on fibre cross-section (cross-section biased mean length) Hauteur H.
3 Mean length based on fibre mass (mass-biased mean length) Barbe B.

The Hauteur and Barbe are the measurements most frequently encountered and are dependent on the method used to measure fibre length.

To see the effect of different fibre diameters on the mean length consider three different fibres each with a different cross-sectional area a and a different length l as shown in Fig. 3.10. The mass (w) of each fibre is therefore $w = a \times l \times \rho$ where ρ is the fibre density.

Mean length L

$$L = \frac{l_1 + l_2 + l_3}{3}$$

In the calculation of mean length each fibre is given an equal weighting no matter how large the diameter of the fibre is.

Cross-section biased mean length H (Hauteur)

In this calculation of mean length each fibre is weighted according to its cross-section, so that if a fibre has a cross-section a_2 which is four times that of a_1 its length will count four times that of a_1 in the calculation of the mean:

$$H = \frac{a_1 l_1 + a_2 l_2 + a_3 l_3}{a_1 + a_2 + a_3}$$

In most cases the percentage of fibres biased by cross-sectional area are approximately equivalent to the percentage of fibres in number so that these diagrams are practically equivalent. The Hauteur is the figure that is automatically produced when capacitance measurements are employed as in the Almeter or the WIRA fibre diagram machine.

Mass-biased mean length B (Barbe)

The Barbe is obtained when the fibre length groups from a comb sorter are each weighed and the average length calculated from the data. The Hauteur can be obtained from the data by dividing the mass of each length group by its length and expressing the result as a percentage:

$$B = \frac{w_1 l_1 + w_2 l_2 + w_3 l_3}{w_1 + w_2 + w_3}$$

The mass is given by $w = a l \rho$.

Therefore if density ρ is assumed constant then:

$$B = \frac{a_1 l_1^2 + a_2 l_2^2 + a_3 l_3^2}{a_1 l_1 + a_2 l_2 + a_3 l_3}$$

The long fibres are thus given a greater weighting than the short ones so that the mean length will be higher than that calculated by the other two methods.

The Barbe is always greater than the Hauteur for a given sample; the two may be interchanged using the following formula [2]:

$$\text{Barbe} = \text{Hauteur} \left(1 + V^2\right)$$

where V is the fractional coefficient of variation of Hauteur, that is coefficient of variation/100. The CV% of length (hauteur) generally lies between 40% and 60% for wool so that, assuming an average value of 50%, the Barbe would be 25% greater than the Hauteur.

3.4   Methods of measurement: direct methods

The problems encountered when measuring fibre length are very similar to those encountered when measuring fibre fineness. In both cases a large number of fibres have to be measured in order to provide a statistically accurate answer. Furthermore any method that involves the handling of individual fibres is very time consuming. The methods used to measure fibre length fall into two main types: the direct measurement of single fibres mainly for research purposes and methods that involve preparing a tuft or bundle of fibres arranged parallel to one another. In this case the fibres can be grouped for measurement or ultimately the measurement can be completely automated.

The simplest direct way of measuring single fibres is by hand. Each end of the fibre is grasped by a pair of tweezers and the fibre stretched alongside a rule. The tension applied when holding the fibres must be just sufficient to remove any crimp but not enough to stretch the fibre. In order to have a more even tension during the measurement a weight may be hung on the end of the fibre but the method then becomes slower still. The British Standard BS 6176 [13] describes the use of a glass plate with a millimetre scale engraved on it. This is smeared with a small amount of liquid paraffin or petroleum jelly and the fibre is stretched along the scale using tweezers. The oil on the glass helps to control the fibres. Alternatively a number of fibres can be mounted on an oiled slide and viewed at a magnification of 5× or 10× using a projector. The length of the fibre, even though it does not follow a straight path, is then measured by an opisometer. These methods, however, are slow and tedious and are used mainly for research.

3.4.1 WIRA fibre length machine

The WIRA fibre length machine [13] is an attempt to automate the process of single fibre measurement and is intended mainly for measuring wool fibres. The equipment shown in Fig. 3.11 involves a rotating shaft with a spiral groove machined in it. One end of the fibre to be measured is gripped by a pair of tweezers whose point is then placed in the moving spiral. This has the effect of moving the tweezers to the right and so steadily drawing the fibre through the pressure plate. This ensures that the fibre is extended under a standard tension. A fine wire rests on the fibre and is arranged so that when the far end of the fibre passes under the wire it allows it to drop

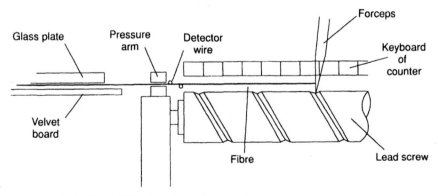

3.11   The WIRA fibre length apparatus.

into a small cup of mercury and thus complete an electrical circuit. This causes the shaft to stop moving, so halting the tweezers; at this point the tweezers are then raised to lift the counter immediately above where it has stopped. The counters are arranged in 0.5 cm sections and each time one is lifted it adds a unit to the appropriate length group so contributing to a cumulative total.

The advantage of the apparatus is that it gives a standard tension to the fibres, it involves less operator fatigue and it gives semi-automatic recording of the results to 0.5 cm intervals. The apparatus is claimed to be able to measure up to 500 fibres per hour. However, the detector wire that is used to sense the end of the fibre is very delicate and it is difficult to set up.

3.5   Methods of measurement: tuft methods

Tuft methods are often used for routine fibre length testing as they are more rapid than the direct methods. The various methods share a common factor in that the preliminary preparation is directed towards producing a bundle of parallel fibres. However, there is always a danger of fibre breakage during the preparation stage when the fibres are changed from being randomly oriented and entangled into being straight and parallel.

3.5.1 Cotton grading

In the simplest form of cotton grading the staple length is estimated by classers who produce a tuft by hand. This is done by holding the fibres in one hand and extracting the fibres by their ends with the other hand. This first step produces a bundle with the fibre ends together. The fibres are then taken from this bundle by hand a few at a time starting with the longest fibres and are laid down next to one another in descending order of length. This produces a fibre diagram as shown in Fig. 3.12.

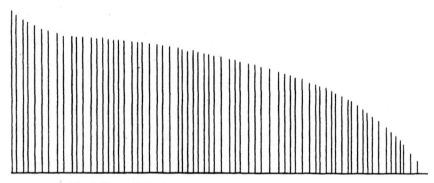

3.12 A fibre diagram.

The classer then chooses limits where there are easily defined edges, ignoring the long and short fibres so that the limits reflect the length of the majority of the fibres. Measurements made in this way depend on the personal skill and judgement of the classer so that they are liable to vary from place to place and over a period of time. To help maintain standards the US Dept. of Agriculture (USDA) has a range of official standards for staple lengths so that the classer can check his or her judgement against these. However, the accuracy is fairly high, staple length being assessed to the nearest $\frac{1}{16}$ inch or $\frac{1}{32}$ inch (1.6 or 0.8 mm) in some cases.

As hand stapling requires experience, alternative methods of tuft measurement have evolved mostly involving similar methods of preparation:

1   Preparation of a fringe or tuft with all fibres aligned at one end.
2   Withdrawal of fibres in order of decreasing length.
3   Preparation of a fibre diagram by laying fibres alongside one another in decreasing order of length with their lower ends in a line.
4   Analysis of the diagram.

3.5.2 Comb sorter

The comb sorter, different versions of which are used for both cotton and wool, uses a set of fine combs arranged at fixed intervals to hold the fibres and keep them straight. Because cotton fibres are comparatively short the cotton comb has two sets of combs, the lower set having the needles pointing upwards and the combs in the upper bed intermeshing with these, the needles pointing downwards. Both the top and bottom combs are set 6 mm ($\frac{1}{4}$ in) apart so that when they are both in place there is a separation of 3 mm ($\frac{1}{8}$ in) between them. A bundle of fibres produced by appropriate sampling procedures is straightened by hand and then pressed into the lower set of combs with the ends of the fibres protruding as shown at A in Fig. 3.13. The end of the bundle is straightened by gripping the ends of the outermost fibres with a wide clamp and withdrawing them a few at a time. The whole sample is then transferred in this way, a few fibres at a time, to position B at the other end of the combs and placed there so that the fibre ends coincide with the first comb.

The sample is pressed down into the bottom combs and the top combs are then lowered onto the sample. The rear combs are moved out of the way one at a time until the ends of the longest fibres are exposed. The exposed fibres are then removed by the grip and laid on a black velvet pad. The next comb is then removed, so exposing the fibres which constitute the next length group and these are removed and laid next to the first set of fibres, making sure that all the fibres are laid with a common base line. The rest of the combs are then dropped in succession so that the fibres

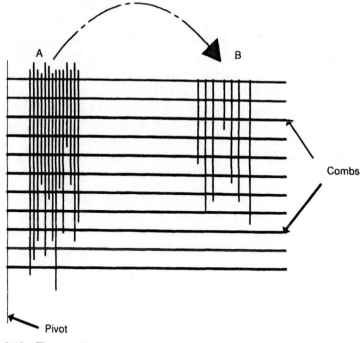

3.13   The comb sorter.

are removed in decreasing order of length, thus giving a comb sorter diagram.

The length values can be computed from a tracing of the fibre diagram, in particular the effective length which is the length of the main bulk of the longer fibres. This ignores the shorter fibres but many machinery settings in the spinning process such as the distance between the drafting rollers are more dependent on the longer fibres.

According to [14] the effective length can be determined from the tracing (Fig. 3.14) by the following construction:

1 Halve the maximum length OA at Q and determine the position of P such that the perpendicular PP′ = OQ.
2 Mark off OK = OP/4 and erect a perpendicular KK′ cutting the curve at K′.
3 Halve KK′ at S and determine the position of R such that the perpendicular R′R = KS.
4 Mark off OL = OR/4 and erect a perpendicular LL′, cutting the curve at L′. The length represented by LL is the effective length.

The effective length is rounded off to the nearest 0.8 mm ($\frac{1}{32}$ in):

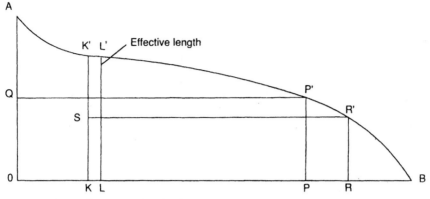

3.14   The analysis of a fibre diagram.

American staple = 0.91 × effective length

The fibres in each length group can be weighed to give a mass histogram.

The US method [15] uses two separate sets of combs, the fibres being transferred from one set to the other rather than from one side of the comb to the other. Before the final array is made the fibres are transferred twice between the combs. When each length group is then removed from the rear of the sample it is kept separate from the other length groups on the velvet board. This allows each length group to be weighed on a sensitive balance (accurate to ± 0.1 mg) so giving the proportion by mass of fibres in each length group. The sample of approximately 75 mg is weighed before the test. The mean length is calculated from the sum of the mass-biased lengths:

$$\text{Mean length} = \frac{\sum WL}{\sum W}$$

where L = group length,
      W = mass of fibre in length group.

The upper quartile length is also calculated. This is the length of fibre at which point a quarter of the fibres by mass are longer than it.

When the length of wool fibres is being measured, the combs in the comb sorter are set at 1 cm intervals but a top set of combs is not used. The sample of fibres in each class is weighed, thus giving a mass distribution.

3.5.3   The clamped tuft method

In this method [16] a suitable length of sliver is hung vertically with a weighted clamp B at the bottom to keep it under tension as shown in

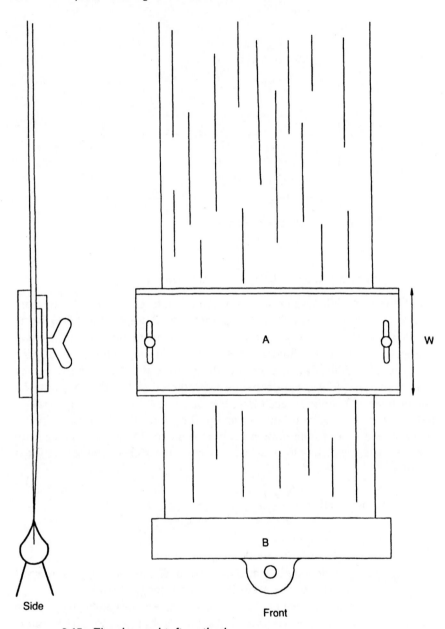

Side   Front

3.15   The clamped tuft method.

Fig. 3.15. Two halves of a metal clamp A are fastened around the sliver so clamping all the fibres that pass through that area. The sliver is then combed at each side of the clamp so that any fibres not gripped are removed, but a tuft is left protruding from each side. The clamp is made with the top narrower than the base so that a blade can be run along each edge cutting off the protruding tufts. The tufts are then weighed together, the clamp opened and its contents weighed separately. The mean fibre length (mass-biased) can then be calculated from the following formula:

$$\text{Mean fibre length} = \frac{W \times \text{total mass of combed tufts}}{\text{mass of clamped sliver}}$$

where W = width of clamp.

3.5.4 Fibrograph

The Fibrograph is an automated method of measuring the fibre length of a cotton sample [17]. It uses an optical method of measuring the density along the length of a tuft of parallel fibres.

The first part of the measuring process is the preparation of a suitable sample. This can be done either by hand or with a Fibrosampler. The Fibrosampler has a rotating brush which withdraws cotton fibres from a perforated drum and deposits them on a comb. The outcome is that the fibres are placed on the comb in such a way that they are caught at random points along their length to form a beard. The sample preparation gives a tuft sample as described in Fig. 1.4. The beard is scanned photoelectrically by the Fibrograph from the base to the tip. The intensity of light that passes through the beard at a given position is used as a measure of the number of fibres that extend to that distance from the comb. The sample density is then plotted against distance from the comb to give a Fibrogram as shown in Fig. 3.16. In the fibrogram random points on the fibres are lined up on the base line and the segments above the base line can be conceived as arrayed in order of length and equally spaced, with the Fibrogram as the envelope of the fibre tips. This method makes the assumption that a fibre is caught on the comb in proportion to its length as compared with the total length of all fibres in the sample and that the point where it is caught is at random along its length. The span lengths at given percentages of fibres are usually measured; the 2.5% span length is considered to correlate with the classer's staple length. From the 50% span length and the 2.5% span length a uniformity index can be calculated:

$$\text{Uniformity index} = \frac{50\% \text{ span length}}{2.5\% \text{ span length}} \times 100$$

3.5.5 WIRA fibre diagram machine

This apparatus is designed to measure the fibre length of combed wool sliver, that is wool in which the fibres have been combed parallel [18].

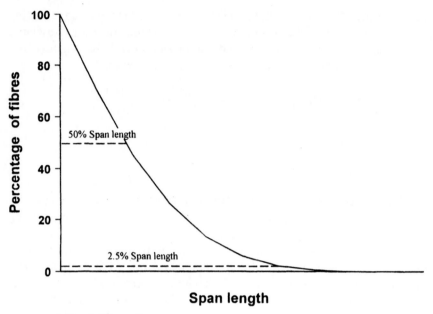

3.16   A Fibrogram.

The method consists of two parts: firstly, the preparation of a fibre sample in which all the fibres have one end sealed between strips of plastic and, secondly, the measurement carried out on the sample. In order to prepare the sample a length of combed sliver is parted by hand and then doubled with the two ends being placed together. This open end is then squared by withdrawing small amounts of the protruding fibres successively with a pair of wide grips. The squared end is then heat sealed between two lengths of polythene tape so that about 3 mm of the fibre ends are held between the tapes. The polythene tapes can then be pulled away from the bulk of the fibres, bringing with them a 'draw' of fibres, all of which have one of their ends sealed between the tapes and which are therefore lined up with one another.

The fibre length distribution is measured by passing this 'draw' thick end first through a measuring slot as shown in Fig. 3.17.

The machine measures the capacitance of the sample as it passes through the slot. The capacitance measurement is proportional to the total amount of material in the slot at that time. The measurement is repeated at known distances along the 'draw' so that a graph can be constructed of the amount of material against the distance from the fibre ends. This graph, which is shown in Fig. 3.18, takes the form of a cumulative length diagram and is similar to the fibre diagram produced by hand (Fig. 3.12).

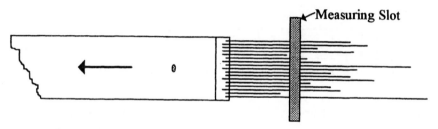

3.17   A draw for the WIRA fibre diagram machine.

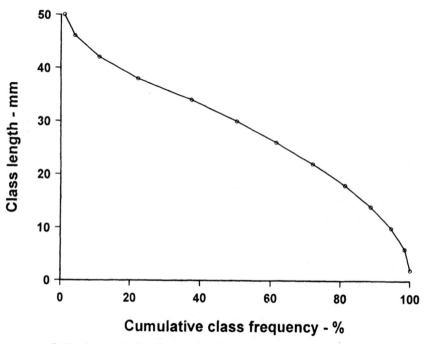

3.18   A cumulative frequency diagram.

The mean length of the fibre sample can be calculated from ten length readings taken at 10% intervals between 5% and 95%. Fibres less than 10mm in length are not measured by this method. As it has been estimated that for wool only 3% of fibres lie below this value, the maximum is therefore set at 97% rather than 100% to allow for this.

3.5.6 Almeter

The Almeter is a capacitance method of measuring fibre length which requires a sample of fibres which are parallel and with one of their ends

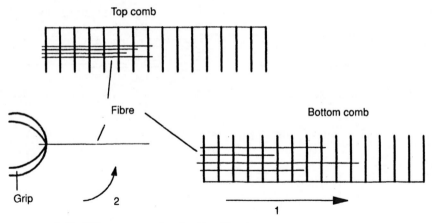

3.19 Fibre preparation for the Almeter.

aligned in a similar way to the WIRA fibre length machine [19]. In this case the sample is prepared by a separate fully automatic machine known as the Fibroliner so that together the two instruments form a system of fibre length measurement not dependent on operator skill.

The Fibroliner (Fig. 3.19) has two sets of combs together with a wide grip which transfers the fibres from one set to the other during the machine cycle. The action of the machine is similar to that taking place during the preparation of a fibre array in a comb sorter. In the case of wool a sample of combed fibre is broken by hand and pressed into the first set of combs so that the fibre ends are facing the grip. The combs are arranged so that at each machine cycle the bed of combs moves forward by one and the front comb drops down out of the way. The combs are moved forward until the ends of the fibres protrude beyond the first comb. The sample can now be squared by operating the grips until a square edge of fibre ends protrudes from the front comb. During each operating cycle the grip removes the fibres by the ends from the first set of combs (1) and they are then transferred mechanically to the second set of combs (2) which are above the grip. The machine is run for sufficient cycles to give the required weight of fibre in the sample, all the fibre draws being placed on top of one another in the second comb with their ends aligned. The top comb is shown removed in Fig. 3.20.

The sample of fibres with aligned ends produced by the Fibroliner is then transferred via the top comb to the sliding sample holder of the Almeter. The top comb is placed upside down over the sample holder in a predetermined position and the fibre sample is pushed out of the combs onto the bottom sheet of plastic film. The top plastic sheet of the sample holder is then lowered onto it, so trapping the sample as shown in Fig. 3.21. The fibres

3.20   The top comb of a Fibroliner.

3.21   A sample in place in the Almeter between plastic sheets.

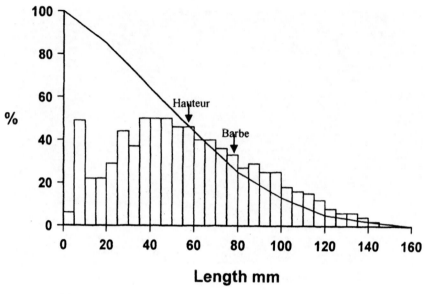

3.22   A cumulative Hauteur diagram and histogram.

are aligned so that the ends are in a line at right angles to the direction of movement of the slide.

During measurement the sample is moved at a constant speed with the longest fibre entering the capacitor first. The change in capacity caused by the presence of the fibres is proportional to the mass of the fibres contained within the width of the capacitor (1.6 mm). Thus the capacity change is proportional to the total cross-sectional area of the fibres. This measurement produces a cumulative Hauteur diagram (Fig. 3.22) similar to that produced by the WIRA fibre diagram machine. From the numerical results the computer can also calculate the Barbe distribution and the mean fibre length for both cases.

3.6   High-volume instruments

With the increasing automation of fibre measurements the trend is to provide a set of instruments linked to a common computer which together can completely characterise a fibre sample and print a report on it. The trend is the most advanced in the case of cotton fibres where the instrument sets are known as high-volume instruments (HVI). The aim of such methods is to analyse every bale of cotton so that a certificate of its quality can be given when it is offered for sale. Existing instruments such as the Fibrograph and the Shirley fineness and maturity tester have been adapted for the measuring lines and new types of instruments have been designed.

One such system is the Spinlab system which measures seven parameters: fibre length, length uniformity, strength, elongation, Micronaire, colour and trash. The colour and trash content are both measured optically. The colour is determined by measuring both the percentage reflectance and the yellowness, which are then combined into a USDA colour grade. The trash is measured using a video camera which counts the number of trash particles in the sample. The Micronaire value is measured in the usual way by airflow but a sample of between 9.5 and 10.5 g is used instead of a fixed weight sample. In this case the actual weight of the sample is entered in the instrument which then makes allowance for it. The remaining fibre properties are all measured by a Fibrograph instrument. The sample for this is prepared by a Fibrosampler which provides a test sample in the form of a beard or tuft clamped by a Fibrocomb. After the loose fibres are brushed out the sample is scanned optically by the Fibrograph to measure the fibre length and length uniformity. After this measurement a set of jaws clamp on the fibre beard and the rear jaw then retracts until the sample is broken. The Pressley or Stelometer equivalent results can be calculated from the breaking force.

The system is capable of testing 180 samples per hour using two operators.

General reading

Morton W E and Hearle J W S, *Physical Properties of Textile Fibres*, 3rd edn. Textile Institute, Manchester, 1993.
Anon, 'The measurement of wool fibre diameter', *Wool Sci Rev*, 1952 **8** 33.

References

1. Matsudaira M, Watt J D and Carnaby G A, 'Measurement of the surface prickle of fabrics Part 1: The evaluation of potential objective methods', *J Text Inst*, 1990 **81** 288–299.
2. The *Wira Textile Data Book*, WIRA Leeds, 1973.
3. BS 2043 Method for the determination of wool fibre fineness by the use of a projection microscope.
4. BS 3183 Method for the determination of wool fibre diameter by the airflow method.
5. IWTO-28-89 Determination by the airflow method of the mean fibre diameter of core samples of raw wool.
6. BS 3181 Part 1 Determination of micronaire value by the single compression airflow method.
7. ASTM D 1448 Micronaire reading of cotton fibers.
8. Lord E, *The origin and assessment of cotton fibre maturity*, Shirley Developments Ltd, Manchester, 1975.

9. Baxter B P, Brims M A and Taylor T B, 'Description and performance of the optical fibre diameter analyser (OFDA)', *J Text Inst*, 1992 **83** 507.

10. Lynch L J and Michie N A, 'An instrument for the rapid automatic measurement of fibre fineness distribution', *Text Res J*, 1976 **46** 653–660.

11. Irvine P A and Lunney H W M, 'Calibration of the CSIRO fibre fineness distribution analyser', *Text Res J*, 1979 **49** 97–101.

12. ASTM D 1577 Linear density of textile fibres.

13. BS 6176 Method for determination of length and length distribution of staple fibres by measurement of single fibres.

14. BS 4044 Methods for determination of fibre length by comb sorter diagram.

15. ASTM D 1440 Length and length distribution of cotton fibres (array method).

16. Anon, 'The measurement of wool fibre length', *Wool Sci Rev*, 1952 **9** 15.

17. ASTM D 1447 Length and length uniformity of cotton fibres by Fibrograph measurement.

18. BS 5182 Measurement of the length of wool fibres processed on the worsted system, using a fibre diagram machine.

19. Grignet J, 'Fibre length', *Wool Sci Rev*, 1980 **56** 81.

4.1 Linear density

The thickness or diameter of a yarn is one of its most fundamental properties. However, it is not possible to measure the diameter of a yarn in any meaningful way. This is because the diameter of a yarn changes quite markedly as it is compressed. Most methods of measuring the diameter of yarn, apart from optical ones, involve compressing the yarn as part of the measurement process. Therefore the measured diameter changes with the pressure used so that there is a need for agreement on the value of pressure at which the yarn diameter is to be defined. On the other hand optical systems of measuring yarn diameter have the problem of defining where the outer edge of the yarn lies as the surface can be rather fuzzy, having many hairs sticking out from it. Therefore the positioning of the yarn boundaries is subject to operator interpretation. Because of these problems a system of denoting the fineness of a yarn by weighing a known length of it has evolved. This quantity is known as the linear density and it can be measured with a high degree of accuracy if a sufficient length of yarn is used.

There are two systems of linear density designation in use: the direct and the indirect.

4.1.1 Direct system

The direct system of denoting linear density is based on measuring the weight per unit length of a yarn. The main systems in use are:

- Tex – weight in grams of 1000 metres
- Decitex – weight in grams of 10,000 metres
- Denier – weight in grams of 9000 metres

1 tex = 10 decitex.

Tex is the preferred SI unit for linear density but it is not yet in common use throughout the textile industry. Other direct systems can be converted

Table 4.1 Multiplying factors for direct systems of yarn linear density [1]

Yarn number system	Symbolic abbreviation	Unit of mass used	Unit of length used	Multiplying factors yarn number to tex value
Tex	T_t	1 g	1 km	
Denier	T_d	1 g	9,000 m	0.1111
Linen, dry spun, Hemp, jute	T_i	1 lb	14,400 yd (spindle unit)	34.45
Woollen (woollen)	Ta_w	1 lb	14,400 yd	34.45

into tex by multiplying by the appropriate factor given in Table 4.1. In the direct system the finer the yarn, the lower is the linear density.

4.1.2 Indirect system

The indirect system is based upon the length per unit weight of a yarn and is usually known as count because it is based on the number of hanks of a certain length which are needed to make up a fixed weight. This is the traditional system of yarn linear density measurement and each branch of the industry has its own system based on the traditional length of hank associated with the locality and the type of yarn manufactured.

The main English ones which are still used every day in the relevant parts of the industry are:

• Yorkshire Skeins Woollen Ny

    Count = number of hanks all 256 yards long in 1 pound

• Worsted Count Ne_w

    Count = number of hanks all 560 yards long in 1 pound

• Cotton Count Ne_c

    Count = number of hanks all 840 yards long in 1 pound

• Metric count N m

    Count = number of kilometre lengths per kilogram

In the indirect systems the finer the yarn, the higher the count.

One way of measuring count is to measure the linear density using the tex system in the first instance and then to convert the result to the

Table 4.2 Constants for conversion of indirect systems of yarn linear density [1]

Yarn count system	Symbolic abbreviation	Unit of length used	Unit of mass used	Constants for conversion to tex values
Cotton bump yarn	N_B	1 yd	1 lb	31,000
Cotton (English)	Ne_c	840 yd (hank)	1 lb	590.5
Linen, wet or dry spun	Ne_L	300 yd (lea)	1 lb	1,654
Metric	Nm	1 km	1 kg	1,000
Spun silk	N_s	840 yd	1 lb	590.5
Typp	Nt	1,000 yd	1 lb	496.1
Woollen (Alloa)	Nal	11,520 yd (spyndle)	24 lb	1,033
Woollen (American cut)	Nac	300 yd	1 lb	1,654
Woollen (American run)	Nar	100 yd	1 oz	310
Woollen (Dewsbury)	Nd	1 yd	1 oz	31,000
Woollen (Galashiels)	Ng	300 yd (cut)	24 oz	2,480
Woollen (Hawick)	Nh	300 yd (cut)	26 oz	2,687
Woollen (Irish)	Ni_w	1 yd	0.25 oz	7,751
Woollen (West of England)	Nwe	320 yd (snap)	1 lb	1,550
Woollen (Yorkshire)	Ny	256 yd (skein)	1 lb	1,938
Worsted	Ne_w	560 yd (hank)	1 lb	885.8

appropriate count system using the appropriate conversion factor K which is given in Table 4.2:

$$\text{Count} = \frac{K}{\text{tex}}$$

4.1.3 Folded yarns

In the traditional count systems a folded yarn is denoted by the count of the singles yarn preceded by a number giving the number of single yarns that make up the folded yarn; for example, 2/24s worsted count implies a yarn made from two 24s count worsted yarns twisted together; 1/12s cotton count means a single 12s count cotton yarn.

In the tex system there are two possible ways of referring to folded yarns: one is based on the linear density of the constituent yarns and the other is based on the resultant linear density of the whole yarn.

In the first way the tex value of the single yarns is followed by a multiplication sign and then the number of single yarns which go to make up the folded yarn, e.g.

$$80 \text{ tex} \times 2$$

This indicates a yarn made from twisting together two 80 tex yarns. This type of designation is generally used with woollen yarns.

In the second way of numbering folded yarns the linear density of the whole yarn is used. This is known as the resultant linear density of the yarn and is preceded by a capital R to denote resultant. This is then followed by an oblique stroke / and the number of single yarns twisted together, e.g.

$$R\ 74\ \text{tex}/2$$

This indicates a yarn made from twisting two yarns together which results in a final yarn whose linear density is 74 tex. This type of designation is generally used with worsted yarns.

The two systems are not identical as there is usually some contraction in length when the single yarns are twisted together so making the resultant count slightly higher than would be expected from the count of the single yarns. Therefore in the above example (R 74 tex/2) the linear density of the individual yarns would have been less than 37 tex.

4.1.4   Measuring linear density

Sampling

For lots that contain five cases or less the sample should consist of all the cases. Ten packages are selected at random but in approximately equal numbers from each case.

For lots that consist of more than five cases, five cases should be selected at random and two packages selected at random from each of these cases. In all cases the sampling ends up with ten packages.

Effect of moisture content

Yarns contain a varying amount of moisture depending on the constituent fibres and the moisture content of the atmosphere where they have been stored. The additional moisture can make an appreciable difference to the weight and hence the linear density of the yarn. Therefore when measuring the linear density of a yarn the moisture content has to be taken into consideration. There are three conventional ways of expressing linear density, each of which has a different way of dealing with the moisture content.

Linear density as received

In this method no allowance is made for the moisture content, the linear density being measured on the yarn as it is. The essence of the method is that a number of skeins are wound on a wrap reel which has a circumference of a convenient length, for example 1 metre. These are then weighed and the linear density calculated from the total length and the weight.

Depending on the linear density and type of fibre used in the construction, yarns can easily be extended by a relatively small load. Therefore when measuring the length of a piece of yarn or when reeling a given length of yarn it is important that the operation is carried out using a standard tension. Because of this factor it is important, for accurate work, that the winding tension used when reeling a hank of yarn on a wrap reel, is correct. The tension on the wrap reel is set by introducing the correct amount of friction into the yarn path. However, the amount of friction introduced is not quantifiable so that the tension has to be set by first making test hanks and then checking their girths on a skein gauge.

The skein gauge which is shown in Fig. 4.1 checks the length of a 50 wrap skein under a standard tension. The test hank is passed round the lower fixed peg and the upper peg which forms one arm of a balance. The load on the other end of the balance is set at 50 g × the nominal tex of the yarn. If the length of the hank is correct the pointer will be opposite the zero mark, any deviation from the correct length is shown directly as a plus or minus percentage. The length of the skein should be within 0.25% of the actual girth of the reel, the reeling tension of the wrap reel being adjusted to achieve this.

Because the yarn on a package may be under tension it is correct practice first to wind a hank from the package of sufficient length for all the tests which are to be carried out. This is then allowed to relax without any tension for 4 h before winding the actual test skeins from it.

Linear density at standard testing atmosphere

In this method the skeins of yarn are preconditioned for 4 h by drying in an oven at 50 °C. They are then conditioned in the standard atmosphere (20 °C and 65% RH) for 24 h. The reason for preconditioning the yarn is so that the equilibrium moisture content is approached from the same side

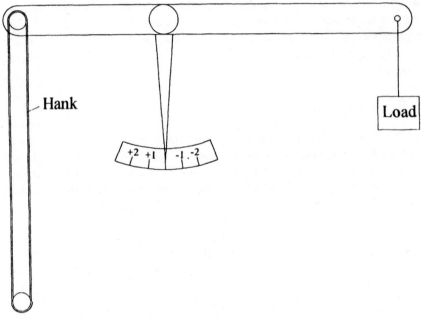

4.1   A skein gauge.

each time, thus avoiding the effects of hysteresis. The reeling of the hanks and calculation of the linear density are then carried out as above.

Linear density at correct condition

This method is more accurate than the previous one as the amount of moisture contained by fibres in equilibrium with the standard atmosphere can vary. In the method the hanks are reeled as above and then dried to oven dry weight (105 °C – two consecutive weighings the same) and weighed. The dry weight then has the appropriate standard regain allowance added to it and the linear density is then calculated from this weight

Weight at correct condition

$$= \text{dry weight} \times \frac{(100 + \text{standard regain})}{100}$$

4.1.5 Linear density from a fabric sample

When the linear density of a yarn has to be determined from a sample of fabric, a strip of the fabric is first cut to a known size. A number of threads are then removed from it and their uncrimped length is determined under

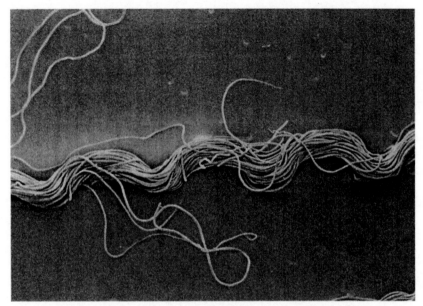

4.2   A portion of yarn removed from a fabric ×24.

a standard tension in a crimp tester. All the threads are weighed together on a sensitive balance and from their total length and total weight the linear density can be calculated.

Yarn from a finished fabric may have had a resin or other type of finish applied to it so that its weight is greater than that of the original yarn. Alternatively it may have lost fibres during the finishing process so that its weight may be lower than that of the original yarn. For these reasons the linear densities of yarn from finished fabrics can only represent an estimate of the linear density of the yarn used to construct the fabric.

Shirley crimp tester

When yarn is removed from a fabric it is no longer straight but it is set into the path that it took in the fabric as shown in Fig. 4.2. This distortion is known as crimp and before the linear density of the yarn can be determined the crimp must be removed and the extended length measured.

The crimp tester is a device for measuring the crimp-free length of a piece of yarn removed from a fabric. The length of the yarn is measured when it is under a standard tension whose value is given in Table 4.3. The instrument is shown diagrammatically in Fig. 4.3 and consists of two clamps, one of which can be slid along a scale and the other which is pivoted so as to apply tension to the yarn. The sample of yarn removed from the fabric is

Table 4.3 Yarn tensions for the crimp tester [2]

Yarn type	Linear density	Tension (cN)
Woollen and worsted	15 to 60 tex	$(0.2 \times tex) + 4$
	61 to 300 tex	$(0.07 \times tex) + 12$
Cotton	7 tex or finer	$0.75 \times tex$
	coarser than 7 tex	$(0.2 \times tex) + 4$
All man-made continuous filament yarn	All	$0.5 \times tex$

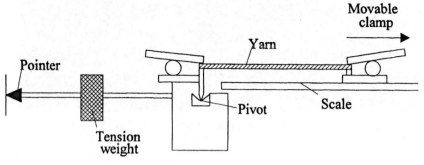

4.3   The Shirley crimp tester.

placed in the clamps with each end a set distance into the clamp. This is because the length of yarn in the clamps has to be allowed for in the measurement. The right hand clamp can be moved along the scale and it has an engraved line on it at which point the extended yarn length can be read. The left hand clamp is balanced on a pivot with a pointer arm attached. On the pointer arm is a weight which can be moved along the arm to change the yarn tension, the set tension being indicated on a scale behind it. At zero tension the left hand clamp assembly is balanced and the pointer arm lines up against a fixed mark. As the weight is moved along the arm the clamp tries to rotate around the pivot, so applying a tension to the yarn.

When a measurement is being made the movable clamp is slid along the scale until the pointer is brought opposite the fixed mark. At this point the tension in the yarn is then the value which was set on the scale. The length of the yarn can then be read off against the engraved line.

The crimp, which is the difference between the extended length and the length of the yarn in the fabric, is defined as:

$$\text{Percentage crimp} = \frac{(L_1 + L_0)}{L_0} \times 100$$

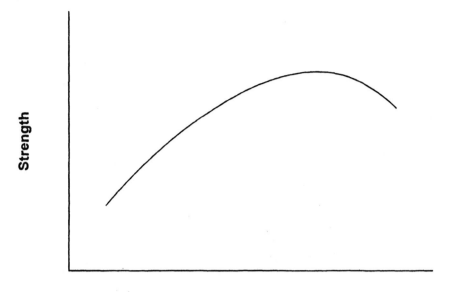

Twist level

4.4   The effect of twist level on strength, staple fibre yarn.

where L_0 = distance between ends of the yarn as it lies in the fabric
L_1 = straightened length of yarn

4.2   Twist

Twist is primarily introduced into a staple yarn in order to hold the constituent fibres together, thus giving strength to the yarn. The effects of the twist are twofold: as the twist increases, the lateral force holding the fibres together is increased so that more of the fibres can contribute to the overall strength of the yarn. Secondly as the twist increases, the angle that the fibres make with the yarn axis increases, so preventing them from developing their maximum strength which occurs when they are oriented in the direction of the applied force. The overall result is that there is a point as twist is increased where the strength of the yarn reaches a maximum value, after which the strength is reduced as the twist is increased still further; this is shown diagrammatically in Fig. 4.4. The twist value required to give the maximum strength to a yarn is generally higher than the twist values in normal use since increased twist also has an effect on other important yarn properties.

A small amount of twist is used in continuous filament yarns to keep the filaments together, but the effect of increasing twist is to reduce the strength

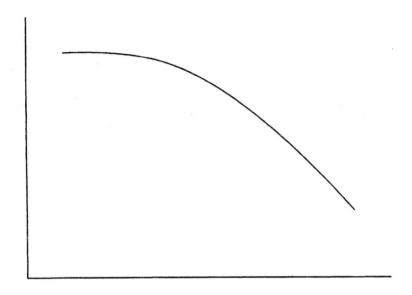

Twist level

4.5   The effect of twist level on strength, continuous filament yarn.

of the yarn below its maximum possible value. The theoretical maximum strength of a continuous filament yarn would be expected to be realised when the filaments were aligned parallel to the yarn axis. However, because of the variability of individual filament strengths the initial effect of twist is to increase the strength of the yarn because of the support given to the weaker filaments. The effect of twist on a continuous filament yarn is shown diagrammatically in Fig. 4.5. The consequence of the effect of twist in reducing the strength of a yarn below its theoretical maximum is that a filament yarn will be stronger than the equivalent staple fibre yarn as a comparatively large amount of twist is always needed in a staple yarn.

The level of twist has other effects on yarn and fabric properties which may override the need for increased strength, including the following:

1   **Handle**. As the twist level in a yarn is increased it becomes more compact because the fibres in it are held more tightly together, so giving a harder feel to the yarn. Furthermore the covering power of the yarn is reduced because of the decrease in the diameter of the yarn. A fabric made from a high-twist yarn will therefore feel harder and will also be thinner. Conversely a fabric produced from a low-twist yarn will have a soft handle which is often a desirable property. However, a reduction in twist level will make the yarn weaker and will also allow the constituent

fibres to be more easily removed with consequences for the pilling and abrasion properties of the fabric.

2 **Moisture absorption**. A high level of twist in a yarn holds the fibres together and hence restricts the access of water to the yarn interior. Therefore such a yarn would be used where a high degree of water repellency is required, for example in a gabardine fabric. A low-twist yarn will absorb water more readily than a high-twist one so would be used in those applications where absorbency is required.

3 **Wearing properties.** The level of twist has effects on two aspects of wear: abrasion and pilling. A high level of twist helps to resist abrasion as the fibres cannot easily be pulled out of the yarn. The same effect also helps to prevent pilling which occurs when fibres are pulled out of the fabric construction and rolled into little balls on the surface.

4 **Aesthetic effects.** The level of twist in a yarn alters its appearance both by changing the thickness and also by altering the light-reflecting properties owing to the change in angle of the fibres. This means that subtle patterns can be produced in a fabric by using similar yarns but with different twist levels. For instance a shadow stripe can be produced by weaving alternate bands of S and Z twist yarns (see Fig. 4.7 below) in the warp. The level of twist can also be used to enhance or subdue a twill effect, depending on whether the fibres in the yarns line up with the twill direction or against it depending on their twist direction.

5 **Faults**. Because the level of twist in a yarn can change its diameter and other properties such as absorption, variation in twist levels in what is nominally the same yarn can change the appearance of a fabric, so giving rise to complaints.

4.2.1 Level of twist

Twist is usually expressed as the number of turns per unit length such as turns per metre or turns per inch. However, the ideal amount of twist varies with the yarn thickness: the thinner the yarn, the greater is the amount of twist that has to be inserted to give the same effect. The factor that determines the effectiveness of the twist is the angle that the fibres make with the yarn axis. In yarns of different linear densities the same angle is produced by different amounts of twist but it leads to the same twist-dependent properties in the yarns. Figure 4.6 shows diagrammatically a fibre taking one full turn of twist in a length of yarn L. The fibre makes an angle θ with the yarn axis. For a given length of yarn the angle is governed by the yarn diameter D:

$$\tan\theta = \frac{\pi D}{L}$$

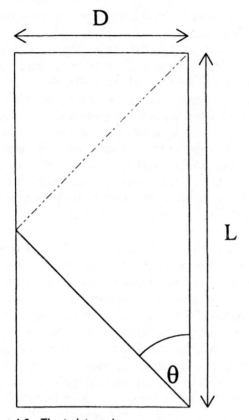

4.6 The twist angle.

The greater the diameter of the yarn, the larger is the angle produced by one turn of twist. As $1/L$ is equivalent to turns per unit length then:

$$\tan\theta \propto D \times \text{turns/unit length}$$

In the indirect system for measuring linear density the diameter is proportional to $1/\sqrt{\text{count}}$. Therefore

$$\tan\theta \propto \frac{\text{turns/unit length}}{\sqrt{\text{count}}}$$

A twist factor is defined using this relationship:

$$K = \frac{\text{turns/unit length}}{\sqrt{\text{count}}}$$

where K is the twist factor.

The numerical value of the twist factor differs with each count system. In the case of direct systems of linear density measurement such as tex:

$$K = \text{turns per metre} \times \sqrt{\text{count}}$$

Typical cotton yarns have twist factors ranging from 3.0 to 8.0 when the measurements are in turns per inch and cotton count (the equivalent values are 29 to 77 when the twist is measured in turns per cm and the linear density in tex). Worsted yarns have twist factors ranging from 1.4 to 2.5 when the twist is measured in turns per inch and the linear density in worsted count (equivalent tex values are 17–29).

A cotton yarn that has a twist factor of 3 will feel soft and docile, whereas a yarn that has a twist factor of 8 will feel hard and lively. A lively yarn is one that twists itself together when it is allowed to hang freely in a loop. Crepe yarns use high twist factors (5.5–8.0 cotton count) to give characteristic decorative effects. A fabric made from such yarns is first wetted and then dried without any constraint. This allows the yarns to curl, producing the characteristic uneven crepe effect.

The twist in a yarn is not usually distributed uniformly along its length. There is a relationship [3] between the twist and the thickness of a yarn which takes the form:

$$\text{Twist} \times \text{mass per unit length} = \text{constant}$$

that is, the twist tends to run into the thin places in a yarn. This means that the twist level will vary along the yarn inversely with the linear density. An uneven yarn will therefore have a twist variability of the same magnitude. Because of this variation it is suggested that the twist level should be determined at fixed intervals along a yarn such as every metre.

4.2.2 Measuring twist

Direction of twist

Twist is conveniently denoted as either S or Z as shown in Fig. 4.7.

Withdrawal of yarn from package

Withdrawal of yarn over the end of a package adds twist to a yarn, whereas withdrawal from the side of the package does not. The British Standard for twist measurement [4] lays down that the yarn should be withdrawn from the package in the manner in which it would be normally used in the next stage of processing. This means that the measured twist may not be the same as the inserted twist.

Twist in yarns, direct counting method

This is the simplest method of twist measurement. It is also the only method recognised as a British Standard [4]; the US standard [5] is similar. The

Z Twist

S Twist

4.7   S and Z twist.

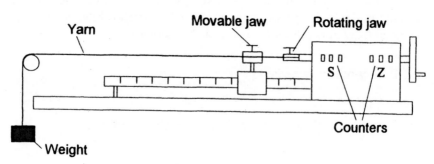

4.8   A simple twist tester.

essence of the method is to unwind the twist in a yarn until the fibres are parallel to the yarn axis and to count how many turns are required to do this. A suitable instrument, an example of which is shown in Fig. 4.8, has two jaws at a set distance apart. One of the jaws is fixed and the other is capable of being rotated. The rotating jaw has a counter attached to it to number the whole turns and fractions of a turn. Before starting any tests the samples should have been conditioned in the standard testing atmosphere.

Testing is started at least one metre from the open end of the yarn as the open end of the yarn is free to untwist so that the level of twist may be lower in that region. As the yarn is being clamped in the instrument it must be kept under a standard tension (0.5 cN/tex) as the length of the yarn will be altered by too high or too low a tension. The twist is removed by turning the rotatable clamp until it is possible to insert a needle between the individual fibres at the non-rotatable clamp end and to traverse the needle across to the rotatable clamp. The use of a magnifying lens may be required in order to test fine yarns.

The twist direction and the mean turns per centimetre or per metre are reported.

Number of tests

1   Single spun yarns. A minimum of 50 tests should be made. The specimen length between clamps must not exceed the average staple length of the yarn. For cotton yarns 10 or 25 mm between the clamps is suitable, for woollen or worsted yarns 25 or 50 mm should be used.

2   Folded and cabled yarns and single continuous filament yarns. A minimum of 20 tests should be carried out with a specimen length of not less than 250 mm.

Continuous twist tester

This apparatus is designed to reduce the amount of handling needed on consecutive twist tests and to speed up the testing process. Yarn passes through the rotating jaw end and is wound up on a rotating drum as it is moved on. Twist is assessed by the same principle as on other twist testers but after removing the twist it is put back into the yarn by rotating the counter back to zero. The rotating clamp is opened and its jaws moved forward to meet the fixed clamp; the jaws are then clamped on the yarn. The fixed clamp is opened and moving jaws are returned to the starting position, taking a new length of yarn with them; the drum takes up the slack in the yarn.

Untwist–twist method

This method is based on the fact that yarns contract in length as the level of twist is increased. Therefore if the twist is subsequently removed, the yarn will increase in length reaching a maximum when all the twist is removed. The method uses a piece of equipment such as that shown in Fig. 4.9 in which the end of the yarn distant from the counter is attached to a pointer which is capable of magnifying its changes in length. At the start of the test the yarn is placed under a suitable tension, either by a clip-on

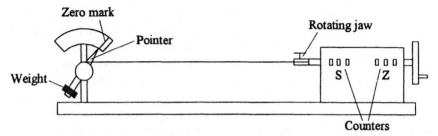

4.9 A tension twist tester.

weight or by a weighted arm as shown. The test procedure is to untwist the yarn until all its twist has been removed and then to continue twisting the yarn in the same direction, until it returns to its original length. The basis of the method is the assumption that the amount of twist put in is equal to the twist that has been removed. However, this is not necessarily the case. For woollen yarns the method may give results up to 20% below the true value, whereas for worsted yarns the results may be 15% higher owing to fibre slippage [6]. One source of error in the method is that at the point of total twist removal the fibres in the yarn are unsupported so that any tension in the yarn may cause the fibres to slip past one another, so increasing the length of the yarn. The difference in length if unnoticed will cause an error in the measurement of turns per unit length. Another source of error is the fact that with some yarns, when the twist is removed, the amount of twist to bring it back to the same length is not equal to the twist taken out.

Because of these problems the method is not recommended for determining the actual twist of a yarn but only for use as a production control method. There is a US standard for this method [7] but it warns that the measured values are only an approximation of the true twist. It suggests that 16 samples are tested using a gauge length of 250 or 125 mm. However, the method is easy to use and has less operator variability than the standard method so that it is often used for measuring the twist in single yarns.

Multiple untwist–twist method

The straightforward untwist–twist method is subject to a variable error owing to the fact that the number of turns to return the yarn to its original length is not the same as the number of turns to take the twist out. This is mainly because when the yarn is spun some of the distortion becomes permanently set into the fibres so that when the twist is removed the yarn is not as straight as it should be. This is particularly a problem in yarns made from wool fibres, especially those that have been deliberately treated in order to set the twist.

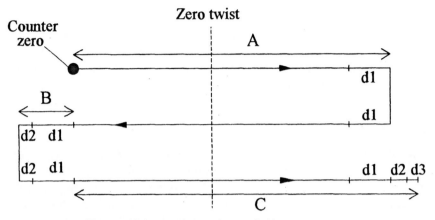

4.10 The multiple untwist—twist method.

The multiple untwist–twist method aims to overcome these problems by repeating the untwisting and twisting action which causes the error due to this source to be progressively reduced. In the test, shown diagrammatically in Fig. 4.10, the yarn is untwisted and retwisted back to its original length as in the normal test and the number of turns A noted. The value A contains an unknown error $d1$. Without the counter being zeroed, the direction of turning is reversed and the yarn untwisted and twisted back to its original length. This ought to bring the yarn back to its original condition, however owing to the errors the counter will show a small number of turns instead of zero. This reading is taken to be B and is due to the errors $d1$ and $d2$. By untwisting and retwisting a third time a further reading C is obtained which contains the errors $d1$, $d2$ and $d3$ as shown. Combining the readings A, B and C gives:

$$A - 2B + C = 4x$$

where x is the number of turns in the length of yarn tested.

The method relies on the errors $d1$, $d2$ and $d3$ becoming progressively smaller so that the remaining error in the above equation is the difference between $d2$ and $d3$ and can be ignored. It is possible to carry out further untwisting and twisting in the same manner to reduce the error even further.

Automatic twist tester

An automatic twist tester is produced (Zweigle D302) which takes the tedium out of the large number of tests required for determining twist. This necessarily depends on untwist–twist type methods for determining twist levels as it cannot detect fibre straightening automatically.

Take-up twist tester

Take-up is the difference between the twisted and untwisted length of a yarn. Twist testers are available with a movable non-rotating jaw which is slid away from the rotating jaw to take up the slack as the twist is removed. This allows the length difference to be measured.

Folded Yarns

Folded or plied yarns have two levels of twist in them. Firstly there is the twist in the individual strands making up the ply and secondly there is the twist that holds the individual plies together. If the twist in the single strands is required the yarns can be analysed by first removing the folding twist and then cutting out individual yarns, leaving the one strand which is to be measured in the twist tester.

4.3   Yarn evenness

Yarn evenness can be defined as the variation in weight per unit length of the yarn or as the variation in its thickness. There are a number of different ways of assessing it.

4.3.1 Visual examination

Yarns to be examined are wrapped onto a matt black surface in equally spaced turns so as to avoid any optical illusions of irregularity. The black-boards are then examined under good lighting conditions using uniform non-directional light. Generally the examination is subjective but the yarn can be compared with a standard if one is available; the ASTM produce a series of cotton yarn appearance standards. Motorised wrapping machines are available: in these the yarn is made to traverse steadily along the board as it is rotated, thus giving a more even spacing. It is preferable to use tapered boards for wrapping the yarn if periodic faults are likely to be present. This is because the yarn may have a repeating fault of a similar spacing to that of one wrap of yarn. By chance it may be hidden behind the board on every turn with a parallel-sided board whereas with a tapered board it will at some point appear on the face.

4.3.2 Cut and weigh methods

This is the simplest way of measuring variation in mass per unit length of a yarn. The method consists of cutting consecutive lengths of the yarn and weighing them. For the method to succeed, however, an accurate way of

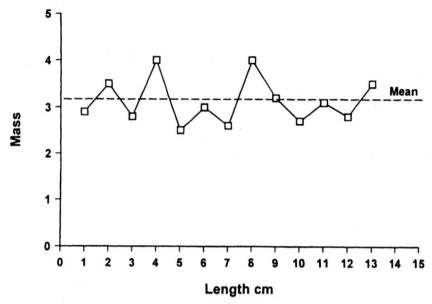

4.11   The variation of weight of consecutive 1 cm lengths of yarn.

cutting the yarn to exactly the same length is required. This is because a small error in measuring the length will cause an equal error in the measured weight in addition to any errors in the weighing operation. One way of achieving accurate cutting to length is to wrap the yarn around a grooved rod which has a circumference of exactly 2.5 cm and then to run a razor blade along the groove, leaving the yarn in equal 2.5 cm lengths. The lengths so produced can then be weighed on a suitable sensitive balance.

If the mass of each consecutive length of yarn is plotted on a graph as in Fig. 4.11, a line showing the mean value can then be drawn on the plot. The scatter of the points about this line will then give a visual indication of the unevenness of the yarn. The further, on average, that the individual points are from the line, the more uneven is the yarn.

A mathematical measure of the unevenness is required which will take account of the distance of the individual points from the mean line and the number of them. There are two main ways of expressing this in use:

1   The average value for all the deviations from the mean is calculated and then expressed as a percentage of the overall mean (percentage mean deviation, PMD). This is termed $U\%$ by the Uster company.

2   The standard deviation is calculated by squaring the deviations from the mean and this is then expressed as a percentage of the overall mean

(coefficient of variation, CV%). This measurement is in accordance with standard statistical procedures.

When the deviations have a normal distribution about the mean the two values are related by the following equation:

$$CV = 1.25 \, PMD$$

4.3.3 Uster evenness tester

The Uster evenness tester measures the thickness variation of a yarn by measuring capacitance [8–10]. The yarn to be assessed is passed through two parallel plates of a capacitor whose value is continuously measured electronically. The presence of the yarn between the plates changes the capacitance of the system which is governed by the mass of material between the plates and its relative permittivity (dielectric constant). If the relative permittivity remains the same then the measurements are directly related to the mass of material between the plates. For the relative permittivity of a yarn to remain the same it must consist of the same type of fibre and its moisture content must be uniform throughout its length. The presence of water in varying amounts or an uneven blend of two or more fibres will alter the relative permittivity in parts of the yarn and hence appear as unevenness.

The unevenness is always expressed as between successive lengths and over a total length. If the successive lengths are short the value is some-times referred to as the short-term unevenness. The measurements made by the Uster instrument are equivalent to weighing successive 1 cm lengths of the yarn.

The measured unevenness arises from various components, the main ones being [11]:

1 The variation in the number of fibres in the yarn cross-section. This is by far the most influential cause of unevenness.
2 In a yarn made from natural fibres the fineness of the fibres themselves is variable leading to a difference in yarn thickness even when the number of fibres in the cross-section remains the same.
3 The inclination of the fibres to the yarn axis can vary. This has the effect of presenting an increased fibre cross-section to the measuring appara-tus. The steeper the angle of inclination of the fibre, the longer is the length that is contained within a fixed length measuring slot.

The Uster tester, besides measuring an overall value of unevenness, also presents a number of other factors derived from the basic measurement of the change in mass along the length of the yarn.

Diagram

A diagram should be plotted of the actual variations in mass per unit length along the length of the yarn.

CV or *U*

The percentage CV or *U* value gives an overall number for yarn irregularity and hence is the most widely used of the measurements that the instrument makes. The *U* value was the only value calculated by the older Uster equipment and is equal to percentage mean deviation. The upper limit of CV which is acceptable for a yarn varies with the different types of yarn. Different spinning systems, counts and end uses have different upper limits and knowledge of these can only be gained from experience of what is acceptable in a given application. Uster produces a volume of 'statistics' which lists the measured values of unevenness for the main types of yarn and for a range of counts for each type, so that measured values can be compared with expected values.

Index of irregularity

There is a natural limit to the evenness that can be achieved in a staple yarn. To produce a completely regular yarn there would need to be exactly the same number of fibres in each cross-section through the yarn. This would mean that the end of one fibre would have to connect with the beginning of the following fibre. No available spinning process can produce such assemblies. The best that can be achieved is complete randomness of the position of individual fibres. On the assumption that all the fibres have the same diameter the theoretical limit to evenness has been calculated as:

$$CV_{lim} = \frac{100}{\sqrt{n}}\%$$

where n = mean number of fibres in the cross-section.

In the case of yarns produced from wool the variations in fibre diameter have also to be taken into account, so that the limiting CV becomes:

$$CV_{lim} = \frac{3.58\,d_f}{\sqrt{T}}$$

where T is the yarn count in tex and d_f is the mean fibre diameter in micrometers. This formula assumes that the coefficient of variation of the wool fibre diameter is 25%.

The formula shows that the finer the count of a yarn, the higher will be its irregularity. This is because when there are only a few fibres in the yarn cross-section the presence or absence of a single fibre makes a bigger difference than if there were a large number of fibres making up the yarn. It is possible to calculate an index of irregularity, I, for any yarn by comparing its measured CV with the theoretical limiting CV.

$$I = \frac{CV_{meas}}{CV_{lim}}$$

To be able to calculate the limiting CV the number of fibres in the cross section of the yarn needs to be known. This number can be calculated from the count of the yarn and the fibre fineness if they are both expressed in the same units. The following formula gives the index of irregularity in terms of the measured CV, the yarn count and fibre fineness.

$$I = \frac{CV_{meas}\sqrt{T}}{100\sqrt{T_f}}$$

where T is the yarn count in tex and T_f is the fibre fineness in tex.

Addition of irregularities

Each machine in the process that produces yarn from fibre adds to the irregularity of the finished yarn. If the irregularities introduced by processes A and B are CV_A and CV_B then the total irregularity can be calculated as follows:

$$CV_{tot} = \sqrt{\left(CV_A^2 + CV_B^2\right)}$$

Imperfections

In addition to measuring the overall variability of yarn thickness the Uster tester also counts the larger short-term deviations from the mean thickness. These are known as imperfections and they comprise thin places, thick places and neps.

The sensitivity of the eye to thick and thin places in a yarn is such that around a 30% change from the mean thickness is needed for a thick or thin place to be visible. In the instrument, therefore, only thick and thin places above these levels are counted. Neps are considered to be those thick places that are shorter than 4 mm whereas areas counted as thick places are the ones that are longer than 4 mm. The total volume of the nep is considered in the assessment but for the purposes of counting they are all assumed to have the same length of 1 mm so that any variation in size is registered as a variation in thickness. Neps are counted at sensitivities of +140%, +200%, +280% and +400% above the mean thickness. For the purpose of the instrument thick and thin places generally have a length equal to the mean fibre length; any places longer than this are considered to be part of the general yarn diameter variation. In general it has been found that the number of imperfections at any one level is related to the imperfections at all other levels so that for comparison purposes it is not important which particular levels are chosen to be recorded.

Spectrogram

An important type of thickness variation is the regular appearance of a thick or thin place at equal intervals along the yarn length. This type of unevenness can give rise to visual effects such as stripiness or moiré patterns in the finished knitted or woven fabric depending on how the repeat length of the fault compares with the fabric width or course length. A level of unevenness which would not be apparent if it was random is much more objectionable if it comes from a regular fault as the eye is very sensitive to pattern.

The spectrogram measures the periodic mass variations in a yarn by analysing the frequencies at which faults occur electronically. From the speed at which the yarn is running the frequencies are converted to wavelengths and slotted into a finite number of discrete wavelength steps. The result is a histogram as shown in Fig. 4.12 where the amplitude is a measure of the number of times a fault of that repeat length occurs. Owing to the fibre length having an effect on the distribution of repeats around that

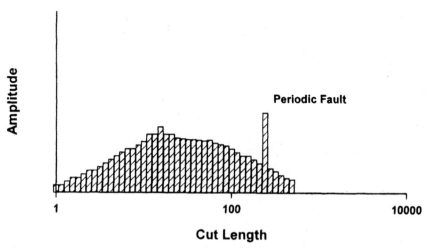

4.12   Spectrogram.

length the background level of the spectrogram is not flat but a periodically repeating fault will show a level much greater than the background as is shown in the figure. As a general rule the height of a peak in the spectrogram should not be more than 50% of the basic spectrogram height at that wavelength.

Theoretical spectrogram

If the CV of a yarn were zero then the spectrogram would consist of a straight line. However, if the yarn has a completely random distribution of staple fibres, as in the case of the limiting CV value, then the staple length L has an effect on the spectrogram.

In the case that all the fibres have the same length then the spectrogram would appear as in Fig. 4.13 with a zero point corresponding to the staple length and a maximum value at 2.7 times the staple length. A diagram of this shape is hardly ever found in practice even in a yarn made from synthetic fibres of constant cut length staple.

In the case of yarns made from natural fibres there is the added complication that the staple length varies quite widely. If L_w is the mean fibre length calculated from the fibre weight staple diagram, the spectrogram then appears similar to that shown in Fig. 4.14 with a less well-defined peak situated at 2.82 times L_w.

The wavelength of the fault gives an indication of its cause and therefore allows it to be traced to such mechanical problems as drafting waves, eccentric or oval rollers in the spinning plant or in earlier preparation stages. The

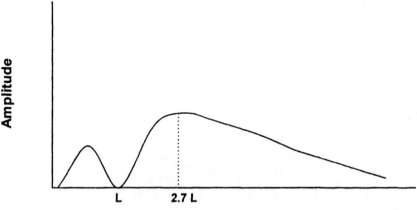

Cut Length

4.13   A theoretical spectrogram for yarn with its staple fibre all the same length L.

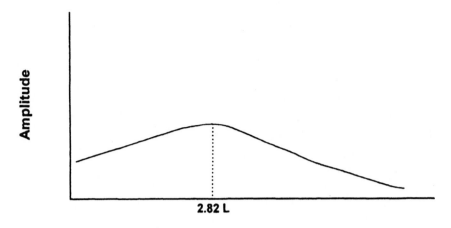

Cut Length

4.14 A theoretical spectrogram for yarn with a variable fibre length; in this case *L* is the mean fibre length.

wavelength can also correspond to the diameter of the yarn package, in which case it will vary between the full and empty package. The wavelength of a fault that occurs before the drafting in the spinning process will be multiplied by the drafting ratio.

Variance length curve

The variance length curve is produced by calculating the CV for different cut lengths and plotting it against the cut length on log–log paper. A perfect yarn would produce a straight line plot. The curve is a useful tool for examining long-term non-periodic variations in a yarn. The better is the evenness of the yarn the lower is the curve and the steeper is the angle it makes to the cut length axis. This is shown in Fig. 4.15 where the variance length curve for an actual cotton yarn is compared with a curve for an ideal yarn.

The measured curve deviates from the theoretical curve in the region where there is long-term variation in the yarn. The variance length curve of a poor fibre assembly lies above the curve of a good fibre assembly as is shown in Fig. 4.16 where the poor yarn diverges from the good yarn at the longer cut lengths.

4.3.4 Zweigle G580

This instrument measures yarn evenness by a fundamentally different method from the mass measuring system of the Uster instrument. Instead

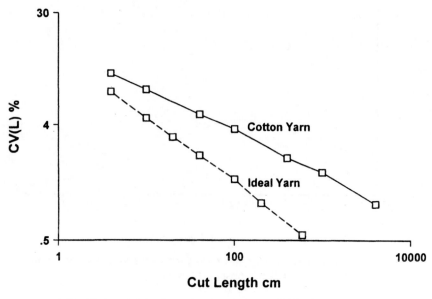

4.15   Variance length curves for cotton and ideal yarns.

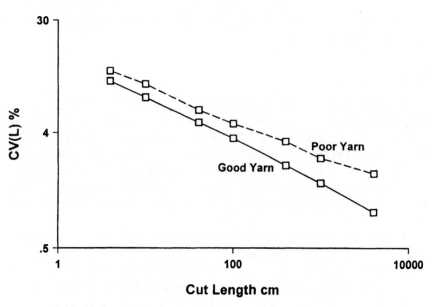

4.16   Variance length curves for poor and good yarns.

of capacitance measurements it uses an optical method of determining the yarn diameter and its variation. In the instrument an infra-red transmitter and two identical receivers are arranged as shown in Fig. 4.17. The yarn passes at speed through one of the beams, blocking a portion of the light

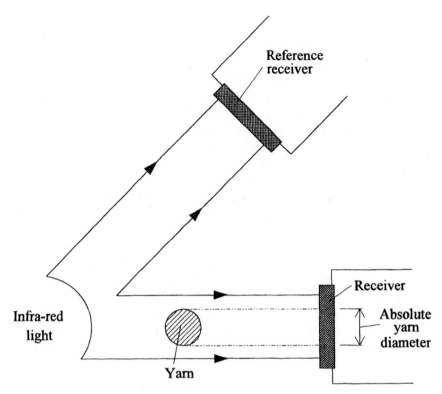

4.17   Zweigle optical evenness.

to the measuring receiver. The intensity of this beam is compared with that measured by the reference receiver and from the difference in intensities a measure of yarn diameter is obtained.

The optical method measures the variations in diameter of a yarn and not in its mass. For a constant level of twist in the yarn the mass of a given length is related to its diameter by the equation:

$$\text{Mass} = CD^2$$

where C = constant,
$\quad D$ = diameter of yarn.

However, in practice the twist level throughout a yarn is not constant [3]. Therefore the imperfections recorded by this instrument differ in nature from those recorded by instruments that measure mass variation. However, the optical system is claimed to be nearer to the human eye in the way that it sees faults. Because of the way yarn evenness is measured, this method is not affected by moisture content or fibre blend variations in the yarn.

4.4   Hairiness

Yarn hairiness is in most circumstances an undesirable property, giving rise to problems in fabric production. Therefore it is important to be able to measure it in order to control it. However, it is not possible to represent hairiness with a single parameter because the number of hairs and the length of hairs both vary independently. For example Fig. 4.18 shows a relatively smooth yarn and Fig. 4.19 a much hairier yarn. Theoretically a yarn may have a small number of long hairs or a large number of short hairs or indeed any combination in between. The problem is then which combination should be given a higher hairiness rating.

It has been found that the number of hairs of different lengths protruding from a yarn is distributed according to an exponential law [12]. That is there are far more short hairs than long ones and the number of hairs falls off exponentially as the hair length increases. It is considered that there are two different exponential mechanisms in operation, one for hairs above 3 mm long and one for those below; this is shown in Fig. 4.20 where the two parts of the plot of the log of the number of hairs against hair length can be approximated by a straight line. The hairiness index devised by [13] assumes a straight line on the plot of the log of the number of hairs against hair length. The number of hairs exceeding 3 mm in length as a percentage of the total number of hairs is found to be

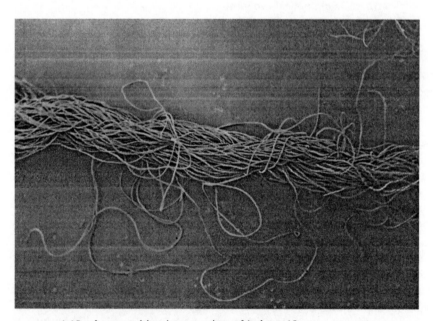

4.18   A yarn with a low number of hairs ×13.

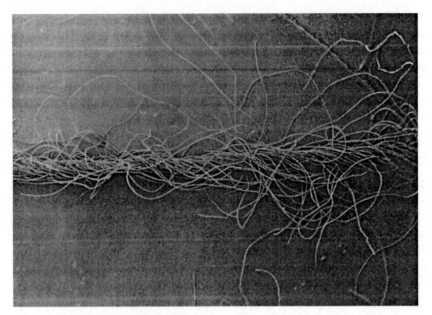

4.19   A hairy yarn ×15.

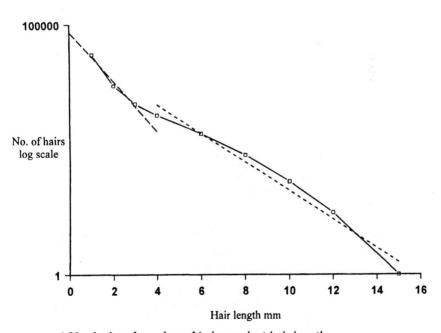

4.20   A plot of number of hairs against hair length.

linearly related to the count of the yarn, that is there are more hairs in a fine yarn than there are in a coarse one of the same type. The overall number of hairs is influenced among other things by the spinning system and by the fibre length.

Measurements of hairiness are very dependent on the experimental configuration used such as the number and type of guides the yarn passes over and also the method chosen for detecting the hairs.

4.4.1 Shirley yarn hairiness tester

The Shirley yarn hairiness tester consists of a light beam shining on a small diameter photoreceptor opposite to it. The yarn under test is run between the light and the receptor at a constant speed. As a hair passes between the light and receptor the light beam is momentarily broken and an electronic circuit counts the interruption as one hair. The instrument has two sets of yarn guides as shown in Fig. 4.21. The lower set leads the yarn over a guide at a fixed distance of 3mm from the receptor. The upper set leads the yarn over a movable guide which can be set at a distance of between 1 and 10mm from the receptor. The total number of hairs in a fixed length of yarn is counted by counting for a given time, the yarn running at a known speed.

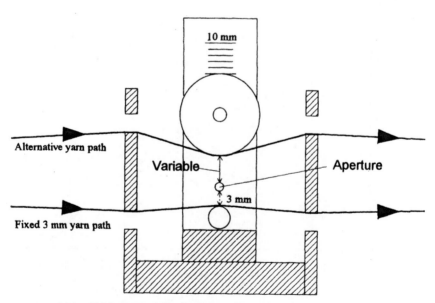

4.21   Shirley yarn hairiness.

4.4.2 Zweigle G565

This apparatus counts the number of hairs at distances from 1 to 25 mm from the yarn edge. The hairs are counted simultaneously by a set of photocells which are arranged at 1, 2, 3, 4, 6, 8, 10, 12, 15, 18, 21 and 25 mm from the yarn as is shown diagrammatically in Fig. 4.22. The yarn is illuminated from the opposite side from the photocells and as the yarn runs past the measuring station the hairs cut the light off momentarily from the photocells, which causes the electrical circuits to count in a similar manner to that of the Shirley instrument. The instrument measures the total number of hairs in each length category for the set test length. The yarn speed is fixed at 50 m/min but the length of yarn tested may be varied. The zero point, that is the position of the yarn edge relative to the photocells, is adjusted while the yarn is running by moving the yarn guides relative to the photocells. A further set of photocells is used to locate the edge of the yarn during the setting up procedure. The instrument calculates the total number of hairs above 3 mm in length which can be used as a comparison with the Shirley instrument. It also computes a hairiness index [13] which has been

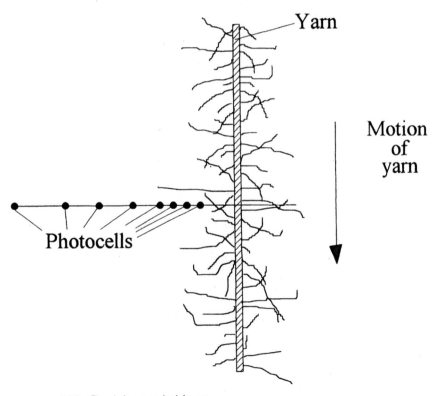

4.22   Zweigle yarn hairiness.

especially devised for this instrument and which is intended to combine all of the information measured by it.

4.4.3 Uster tester 3 hairiness meter attachment

This device is produced as an attachment for the Uster evenness tester and is connected in place of the normal measuring capacitor. However, it makes use of the full statistical result collection capabilities of the evenness instrument. The principle of the measurement is quite different from the above instruments and therefore the results from the two types of instrument are not comparable. In this instrument the yarn is illuminated by a parallel beam of infra-red light as it runs through the measuring head. Only the light that is scattered by fibres protruding from the main body of the yarn reaches the detector as is shown in Fig. 4.23. The direct light is blocked from reaching the detector by an opaque stop. The amount of scattered light is then a measure of hairiness and it is converted to an electrical signal by the apparatus. The instrument is thus monitoring only total hairiness, but using the Uster evenness data collection system can monitor changes in hairiness along the yarn by means of a diagram, spectrogram, CV of hairiness, and mean hairiness in a manner similar to that used in evenness testing.

4.5 Yarn bulk

The WRONZ Bulkometer test gives an indication of the covering power of a yarn when it is incorporated into finished products such as knitwear or

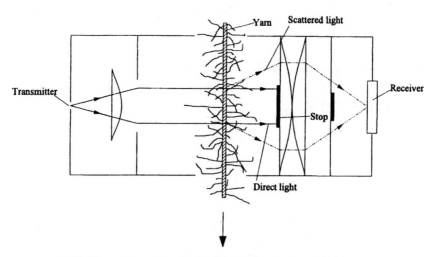

4.23   The measurement of hairiness by scattered light.

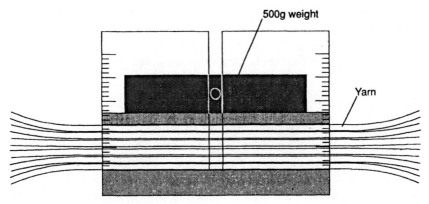

500g weight

Yarn

4.24 The Wronz bulkometer.

carpets. Yarn bulk is defined for the purpose of this test as the volume occupied by 1 g of yarn at a given pressure, measured in cm^3/g.

To carry out the test a hank of yarn containing a known number of turns is placed in a channel 10 cm long by 5 cm wide so that all the strands of the hank are aligned as shown in Fig. 4.24. A load of 500 g is then placed on the sample, so compressing the yarn. When the load comes to rest the height of it above the base is measured. From the area of the channel ($50 cm^2$) and the height of the load, the volume occupied by the yarn can then be calculated.

The size of hank used in the test depends on the linear density of the yarn; a suitable size can be calculated from the formula:

$$\text{Number of turns} = \frac{90,000}{\text{linear density in tex}}$$

It is preferable when comparing similar yarns to keep to the same number of turns.

4.5.1 Textured filament yarns

A large amount of continuous filament yarn has its bulk and stretch increased by some form of crimping process so that it may have the same covering power as staple fibre yarn and approximately the same texture. Tests for yarn stretch, which is related to yarn bulk, usually measure the difference in length between the straightened yarn and the contracted yarn. In order to do this, firstly a load sufficiently heavy to remove the major part of the crimp from the yarn and to straighten the yarn filaments is applied and the yarn measured. This load is removed and then a second load is applied which is sufficiently light to allow the crimp to develop but to keep the yarn straight enough in order to measure it. Some yarns have the crimp set in by the yarn producer and some have a latent crimp which is developed by heating in steam or hot water by the end user. These latter yarns have to be treated to bulk them at some point in the test.

HATRA crimp rigidity

In this test a hank of the yarn is wound under tension sufficiently high as to remove the crimp, the number of turns on the hank being governed by the yarn denier. Two weights are hung on this hank, firstly a small S-shaped weight (0.002 g/den) and on the end of this a much larger weight (0.1 g/den). The sizes of the weights are both governed by the total denier of the hank so that the required tensions are maintained. The whole assembly of hank with two weights is then immersed in a tall cylinder of water. An adjustable stretch rubber rule marked from 0 to 100% is adjusted so that the 100% mark is level with the top of the hank and the 0% mark is level with the bottom of the hank.

After 2 min immersion the bottom heavy weight is removed, leaving the small weight in place but allowing the yarn to contract. After a further 2 min the percentage contraction or crimp rigidity is read directly from the scale.

4.6   Friction

A yarn which is being knitted or woven into a fabric or wound onto a package runs around many guides during the process. Each one causes a drag on the yarn due to friction. Changes in the frictional properties of the yarn can cause an increase or decrease in this drag and hence the tension in the yarn. This can give rise to problems in that too much or too little yarn is fed to a process or the yarn is too tight or too slack. Hence the frictional properties of a yarn are important for its smooth running on production machinery.

When an attempt is made to slide an object resting on a surface a force is required to start the object moving. Once the object is moving, the force required to keep it moving is lower than the original starting force. The force that resists the movement of an object in any direction is known as the frictional force. The force that has to be overcome in order to initiate movement is known as the limiting friction (static friction) and the frictional force that opposes movement when the object is in motion is known as sliding or dynamic friction.

The frictional force is governed by two main factors: the nature of the surfaces in contact and the force that holds the surfaces in contact, which is known as the normal force. The phenomenon of friction is governed by a number of 'laws' which are often known as Amonton's laws:

1   The limiting frictional force F_L is proportional to the normal force R between the surfaces at right angles to the plane of contact. The ratio F_L/R is called the coefficient of static friction μ_L

$$F_L = \mu_L R$$

2   With all ordinary surfaces the limiting friction is independent of the area of contact for a constant normal force.

3   When motion occurs the sliding frictional force F is proportional to the normal force R between the surfaces. The ratio F/R is called the coefficient of sliding friction μ

$$F = \mu R$$

4   With all ordinary surfaces the sliding friction is independent of the area of contact and also independent of the speed of motion within limits.

These laws hold fairly well for hard materials, but not for textile materials particularly at low values of normal force. In most cases it is the sliding friction that is of practical interest.

4.6.1 Coil friction

The friction that a yarn or similar object experiences when running over a curved surface is governed by the angle of contact with the surface and the tension at either side of the contact. This is shown diagrammatically in Fig. 4.25. The tension on the uptake side T_1 is always higher than on the feed side T_2 as the motion of the yarn is resisted by the frictional force:

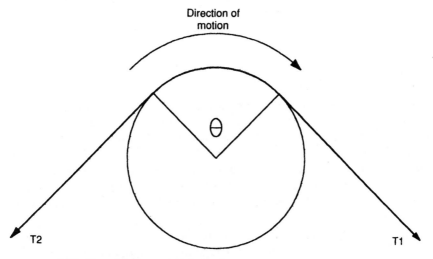

4.25   Yarn friction around a rod.

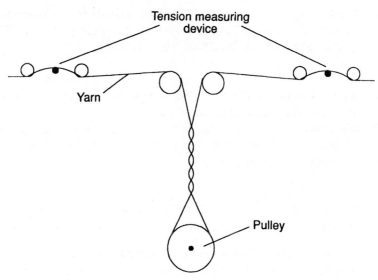

4.26 Yarn to yarn friction.

$$\frac{T_1}{T_2} = e^{\mu\theta}$$

$$\mu = \frac{1}{\theta}(\ln T_1 - \ln T_2)$$

where μ = coil friction,

θ = angular contact in radians.

The frictional force increases rapidly with the angle of contact owing to an increase in the normal force rather than to the increased area of contact. If the angle of contact is kept constant by using a rod of larger radius then the frictional force remains the same although the area of contact has increased.

4.6.2 Measuring yarn friction

The more usual way of measuring yarn friction is to run it around a solid rod and measure the T_1 and T_2 making use of the above relationship. The problem is that the frictional force does not conform closely to this relationship [14] but depends on factors such as the diameter of the rod, the angle of wrap, the input tension and the running speed. Therefore the measurement of the coefficient of friction of a yarn is only of use for comparative purposes when all other factors apart from the yarn under test are kept constant.

The US standard [15] suggests the following test conditions: speed of yarn 100 m/min, either 180° or 360° wrap angle but not less than 90°. The stan-

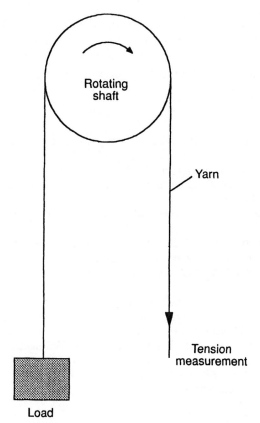

4.27   Yarn to yarn friction: capstan method.

dard friction surface is 12.7 mm (0.5 in) diameter chrome-plated steel of 4–6 μm roughness. It is important that the friction element and other contact areas are cleaned with solvent before each test.

The coefficient of friction is calculated from the measured input and output tensions as the yarn runs at constant speed over the rod.

Yarn friction can also be measured as a yarn to yarn friction instead of yarn to metal. Reference [16] describes two methods of achieving this. One method is to replace the metal rod in the above test with a free-running pulley and to twist the yarn through three complete revolutions as shown in Fig. 4.26 so that the yarn is twisted around itself. The wrap angle is 18.85 rad (1080 deg). The input tension should be set at 1 gf/tex (9.81 mN/tex) and the speed at 0.02 m/min.

The other method is the capstan method in which yarn is wrapped around a tube of 48 mm (1.75 in) diameter to form a covering. A separate strand of yarn is hung over the tube with a weight on one end to give a tension of 1 gf/tex (9.81 mN/tex) and a tension measuring device on the other as shown in Fig. 4.27. The tube is then rotated to give a surface speed of 0.02 m/min and from the measured tension the coefficient of friction can be calculated in a similar manner to that used in the other yarn friction tests.

References

1. BS 947 Specification for a universal system for designating linear density of textiles (tex system).
2. BS 2863 Method for determination of crimp of yarn in fabric.
3. Sust A and Barella A, 'Twist, diameter, and unevenness of yarns a new approach', *J Text Inst*, 1964 **55** T1.
4. BS 2085 Method of test for determination of twist in yarns, direct counting method.
5. ASTM D 1423 Test method for twist in yarns by the direct counting method.
6. Booth J E, *Principles of Textile Testing*, 3rd Edn, Butterworth, London, 1968.
7. ASTM D 1422 Test method for twist in single spun yarns by the untwist–retwist method.
8. Slater K, 'Yarn evenness', *Text Progress*, 1986 **14**.
9. Furter R, *Evenness Testing in Yarn Production*, Parts 1 and 2, Textile Institute and Zellweger Uster AG, Manchester, 1982.
10. ASTM D 1425 Test method for unevenness of textile strands using Zellweger Uster capacitance testing equipment.
11. Zeidman M I, Suh M W and Batra S K, 'A new perspective on yarn unevenness: components and determinants of general unevenness', *Text Res J*, 1990 **60** 1.
12. Barella A and Manich A M, 'The hair length distribution of yarns measured by means of the Zweigle G 565 hairiness meter', *J Text Inst*, 1993 **84** 326.
13. Mangold G and Topf W, 'Hairiness and hairiness index, a new measuring method', *Melliand Textilber*, 1985 **66** 245–247.
14. Rubenstein C, 'The friction of a yarn lapping a cylinder', *J Text Inst*, 1958 **49** T181.
15. ASTM D 3108 Test method for coefficient of friction, yarn to metal.
16. ASTM D 3412 Test method for coefficient of friction, yarn to yarn.

Strength and elongation tests

5.1   Introduction

The level of strength required from a yarn or fabric depends on its end use. For some end uses it is the case that the higher the strength of the materials, the better it is for its end use. This is particularly true for yarns and fabrics intended for industrial products. However, fabrics intended for household or apparel use merely need an adequate strength in order to withstand handling during production and use. It is generally the case that a higher-strength product can only be obtained by either making a heavier, stiffer fabric or by using synthetic fibres in place of natural ones. In either case changes are produced in other properties of the material, such as the stiffness and handle, which may not be desirable for a particular end use.

5.2   Definitions

See also reference [1].

5.2.1   Units

It is important when measuring strength to be clear about the distinction among mass, weight and force. The mass of a body is the term used to denote the quantity of matter it contains, it is a fixed property of an object and does not depend on where it happens to be. The SI unit of mass is the kilogram (kg).

Force can only be defined in terms of what it does. Force is that which changes a body's state of rest or of uniform motion in a straight line. In other words a force causes a body to accelerate. The SI unit of force the newton (N) is defined in terms of the acceleration produced when the force acts on a mass of one kilogram. A newton is defined as the force that when applied to a mass of one kilogram gives it an acceleration of one metre per second per second. In a strength test the result should be measured in units of force rather than units of mass.

Gravitational force pulls all bodies towards the Earth. If a mass of one kilogram is allowed to fall freely in a vacuum towards the Earth it acquires an acceleration of about 9.8 m/s². Using the definition of the newton this implies that the force acting on it must be 9.8 N. If the kilogram is resting on the Earth's surface it will press down on the surface with a force of 9.8 N. This force is what is known as the weight of a body and it is the quantity that is measured by a spring balance. However, the force of gravity on a body varies slightly from place to place on the earth, consequently the weight of an object changes, which is why units of force are preferred to weights for strength measurements.

5.2.2 Breaking strength; tensile strength

This is the maximum tensile force recorded in extending a test piece to breaking point. It is the figure that is generally referred to as strength. The force at which a specimen breaks is directly proportional to its cross-sectional area, therefore when comparing the strengths of different fibres, yarns and fabrics allowances have to be made for this.

The tensile force recorded at the moment of rupture is sometimes referred to as the tensile strength at break [1]. This figure may be different from the tensile strength defined above as the elongation of the specimen may continue after the maximum tensile force has been developed as shown in Fig. 5.1 so that the tensile strength at break is lower than the tensile strength.

5.2.3 Stress

Stress is a way of expressing the force on a material in a way that allows for the effect of the cross-sectional area of the specimen on the force needed to break it:

$$\text{Stress} = \frac{\text{force applied}}{\text{cross-sectional area}}$$

In the case of textile materials the cross-sectional area can only be easily measured in the case of fibres with circular cross-sections. The cross-sections of yarns and fabrics contain an unknown amount of space as well as fibres so that in these cases the cross-sectional area is not clearly defined. Therefore stress is only used in a limited number of applications involving fibres.

5.2.4 Specific (mass) stress

Specific stress is a more useful measurement of stress in the case of yarns as their cross-sectional area is not known. The linear density of the yarn is

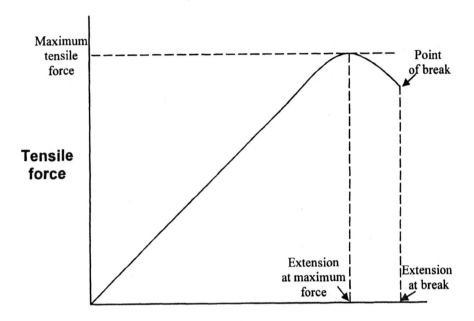

5.1   A force extension curve.

used instead of the cross-sectional area as a measure of yarn thickness. This allows the strengths of yarns of different linear densities to be compared. It is defined as the ratio of force to the linear density:

$$\text{Specific stress} = \frac{\text{force}}{\text{linear density}}$$

The preferred units are N/tex or mN/tex, other units which are found in the industry are: gf/denier and cN/dtex.

5.2.5 Tenacity

Tenacity is defined as the specific stress corresponding with the maximum force on a force/extension curve. The nominal denier or tex of the yarn or fibre is the figure used in the calculation; no allowance is made for any thinning of the specimen as it elongates.

5.2.6 Breaking length

Breaking length is an older measure of tenacity and is defined as the theoretical length of a specimen of yarn whose weight would exert a force sufficient to break the specimen. It is usually measured in kilometres.

5.2.7 Elongation

Elongation is the increase in length of the specimen from its starting length expressed in units of length. The distance that a material will extend under a given force is proportional to its original length, therefore elongation is usually quoted as strain or percentage extension. The elongation at the maximum force is the figure most often quoted.

5.2.8 Strain

The elongation that a specimen undergoes is proportional to its initial length. Strain expresses the elongation as a fraction of the original length:

$$\text{Strain} = \frac{\text{elongation}}{\text{initial length}}$$

5.2.9 Extension percentage

This measure is the strain expressed as a percentage rather than a fraction

$$\text{Extension} = \frac{\text{elongation}}{\text{initial length}} \times 100\%$$

Breaking extension is the extension percentage at the breaking point.

5.2.10 Gauge length

The gauge length is the original length of that portion of the specimen over which the strain or change of length is determined.

5.3   Force elongation curve

When an increasing force is gradually applied to a textile material so that it extends and eventually breaks, the plot of the applied force against the amount that the specimen extends is known as a force–elongation or stress–strain curve. The curve contains far more information than just the tensile strength of the material. The principal features of a force elongation curve, in this case of a wool fibre, are shown in Fig. 5.2. The use of the force elongation curve as a whole allows a better comparison of textile materials to be made as it contains more information about the behaviour of the material under stress than do the simple figures for tensile strength and elongation.

The most important features of the curve are as follows.

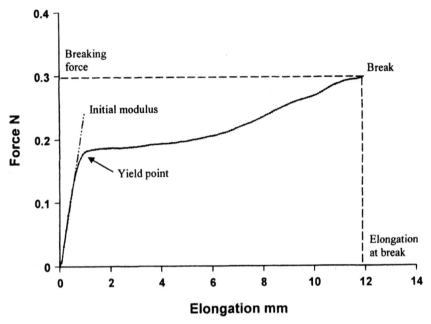

5.2   A typical force elongation curve.

5.3.1 Yield point

Depending on the material being tested, the curve often contains a point where a marked decrease in slope occurs. This point is known as the yield point. At this point important changes in the force elongation relationship occur. Before the yield point the extension of the material is considered to be elastic, that is the sample will revert to its original length when the force is removed. Above the yield point in most fibres, some of the extension is non-recoverable, that is the sample retains some of its extension when the force is removed. This is an over-simplification as in practice there is no clear demarcation between elastic and non-elastic behaviour of textile materials as not all the extension is recoverable even in the elastic region. The change in properties is seen at its most marked in undrawn or partially oriented material (Fig 5.3) because at this point the orientation of the polymer molecules is improved and the material is said to 'draw'. The material continues to extend without an increase in the applied force until it reaches a limit. When the material has finished drawing, the rest of the force extension curve represents the properties of the drawn material which has a higher tenacity owing to the improved orientation.

The yield point is not a definite point on the curve; more often there is a region of continuous change in the curvature between the two different

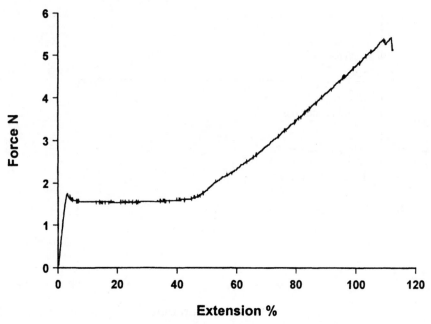

5.3   A force extension curve for partially oriented yarn.

parts of the curve. Therefore in order to measure the force or extension at which it occurs it is necessary to define it. There are four possible methods of defining the yield point.

Slope threshold

In this method the slope of the initial linear region is determined, shown as A in Fig. 5.4 and the point where the slope of the curve decreases to a specified fraction of the initial slope shown at B is taken as the yield point.

Offset yield

In this method the slope of the initial linear region is determined as before. The offset yield point is then defined as the point on the curve where a line parallel to the initial linear modulus region of the curve, but offset from it by a definite value of extension, intersects the test curve. This is shown in Fig. 5.5.

Zero slope method

In this method the point at which the slope of the test curve falls to zero is found. This method is only applicable to certain specimens such as that shown in Fig. 5.6 where the slope actually falls to zero.

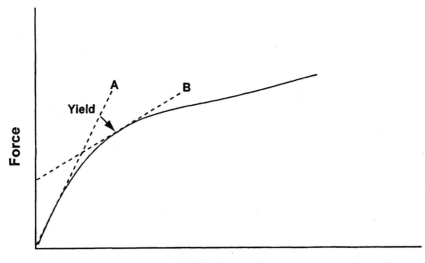

Elongation

5.4   Yield point by the slope threshold method.

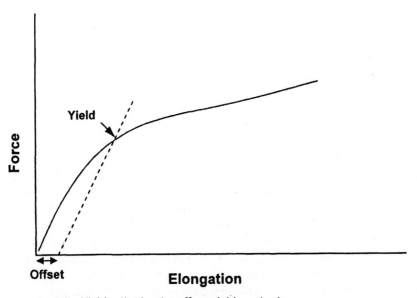

Offset   **Elongation**

5.5   Yield point by the offset yield method.

Meredith's construction

In this method suggested by Meredith [2] the line joining the origin with the breaking point is first constructed. The yield point is then defined as the point on the curve at which the tangent is parallel to this line as shown in

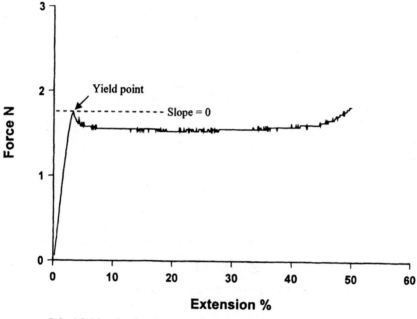

5.6   Yield point by the zero slope method.

Fig. 5.7. This is not an easy method to implement automatically with commercial software packages.

5.3.2 Modulus

The slope of the first linear part of the curve up to the yield point is known as the initial modulus (Young's modulus) and it is the value generally referred to when speaking of modulus without qualification. Modulus as a general term means the slope of the force elongation curve and it is a measure of the stiffness of the material, that is its resistance to extension. The higher the modulus of a material, the less it extends for a given force. If the curve is plotted in terms of stress against strain the units of modulus are the same as those of stress, that is force per unit area such as pascals. If the curve is plotted in terms of force against elongation the units of modulus are those of force/elongation and they depend on whether the elongation is measured in distance, percentage extension or strain.

The use of computer software to record and analyse force elongation curves means that the ways of specifying the modulus of a curve have to be clarified. It is no longer possible to lay a rule on the curve and judge the best position by eye. There are a number of possible moduli that may be measured.

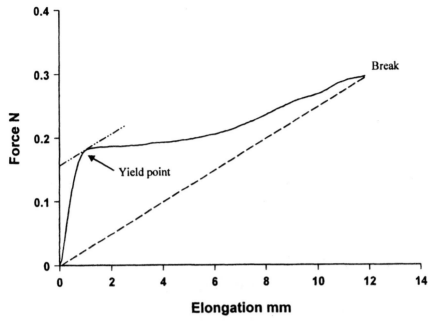

5.7   Meredith's construction for yield point.

Young's modulus

This value is obtained from the slope of the least squares fit straight line made through the steepest linear region of the curve as shown in Fig. 5.8.

Chord modulus

This value is the slope of the straight line drawn between two specified points on the curve as shown in Fig. 5.9. It is not necessary to know the details of the curve between the two points as the value can be derived from measurements of the difference in force between two given values of extension or the difference in extension between two given values of force.

Secant modulus

This value is the slope of the straight line drawn between zero and a specified point on the curve as shown in Fig. 5.10. It is often measured simply as the value of extension at a given force or alternatively as the value of force at a given extension. As a measurement it is sensitive to the amount of extension given to the sample as it is being loaded into the clamps.

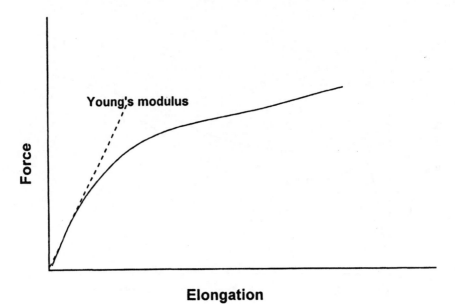

5.8 Young's modulus.

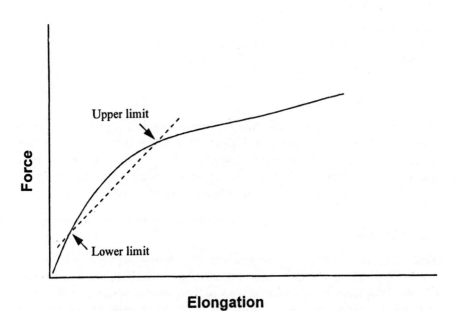

5.9 Chord modulus.

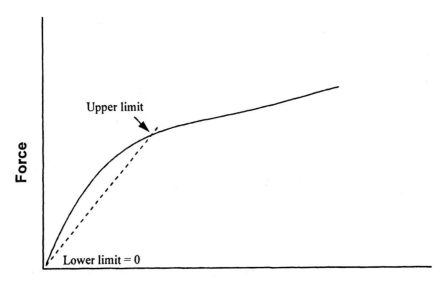

Elongation

5.10 Secant modulus.

Tangent modulus

This value is the slope of the straight line drawn at a tangent to the curve at a specified point as shown in Fig. 5.11. This is the mathematically correct value for the slope of a continuously changing curve. It is not easy to obtain by simple methods but can be readily measured by computer software.

5.3.3 Work of rupture

The work of rupture is a measure of the toughness of a material as it is the total energy required to break the material. Consider a small section of the force extension curve as shown in Fig. 5.12. Within this small section the force can be considered to be constant at a value F. This force increases the sample in length by an amount dl, therefore

Work done = force × displacement = Fdl

From this the total work done in breaking the material which is the work of rupture is:

$$\text{Work of rupture} = \int_0^{\text{break}} F \, dl$$

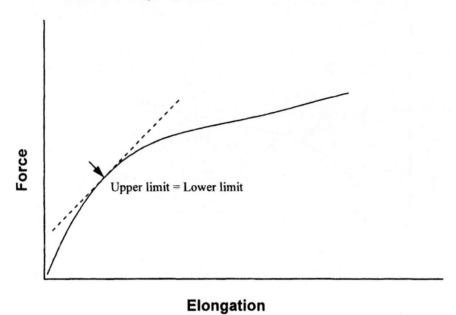

Elongation

5.11   Tangent modulus.

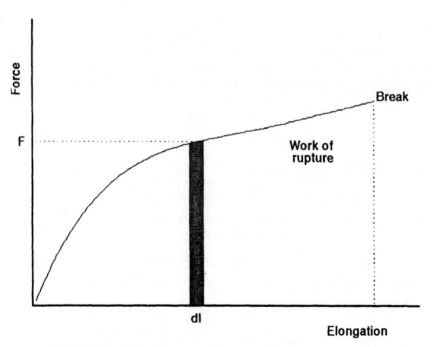

5.12   Work of rupture.

This integral is equivalent to the area under the force extension curve. The SI unit of work is the joule.

The work of rupture of a material is proportional to its cross-sectional area or, more conveniently for yarns and fibres, their linear density as the breaking load is proportional to this. The work of rupture is also proportional to the original length of the material as the elongation of the material is dependent on this. Therefore in order to compare materials it may be necessary to use the specific work of rupture:

$$\text{Specific work of rupture} = \frac{\text{work of rupture}}{(\text{mass/unit length}) \times \text{initial length}}$$

For comparative purposes when the tests use a common gauge length the initial length may be omitted from this formula.

The work of rupture gives a measure of the ability of the material to withstand sudden shocks of a given energy. If a mass m is attached to a thread and dropped from a height h it will acquire a kinetic energy equal to mgh. The thread is capable of withstanding the shock of the fall if its work of rupture is greater than mgh. If the work of rupture is less than this the thread will break. The capacity of a textile material to absorb energy is obviously useful in such applications as car seatbelts or climbing ropes where the ability to safely slow down a moving body is important. It also has importance in other areas which are not so obvious such as tearing resistance or abrasion resistance where high-energy absorption improves these properties. Figure 5.13 shows the force elongation curves of two fabrics of similar tensile strength. The cotton fabric has a slightly higher strength but a work of rupture of 2.36 J whereas the wool fabric has a work of rupture of 8.02 J.

5.3.4 Time dependence

Even in the initial straight line region of the force extension curve textile materials do not behave as strictly elastic materials. Their behaviour is fitted better by a viscoelastic model [3] as the relationship between applied stress and resultant strain contains a time-dependent element. This implies that when the material is extended by an applied force there is, besides the elastic component, a further component whose action opposes the applied force but whose magnitude depends on the speed of extension. This second component decays relatively slowly with time. When the applied force is subsequently removed, the same component also acts to resist the internal elastic forces that bring about contraction.

This time dependence is seen when a yarn or fabric is extended by a given amount and then held at that extended length. If the force required to do this is monitored, it is found to rise immediately to a maximum value and

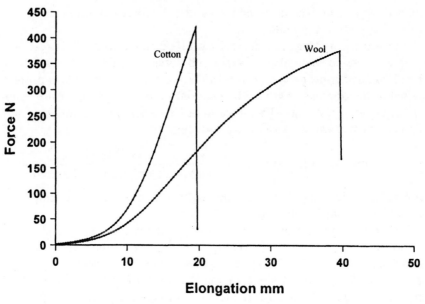

5.13 Two fabrics with different work of ruptures.

then slowly decrease with the passage of time as is shown in Fig. 5.14; this phenomenon is known as stress relaxation.

If instead of a fixed extension, a fixed force is applied to the material, there is found to be an initial extension of a magnitude that is expected from the force extension curve followed by a further slow extension with time. This behaviour is known as creep and its magnitude is an important property to consider when assessing materials that have to be kept under load for a long period of time, such as geotextiles. As the level of force, usually expressed as a percentage of the tensile strength, increases, the rate of creep increases. The rate of creep also increases with increasing temperature. Depending on the level of force and the type of fibre, creep can continue indefinitely until the material fails. The level of force should be set so that the time to failure is longer than the expected life of the product.

After a textile material has been subject to a force even for a short period of time the complete removal of the force allows the specimen to recover its original dimensions, rapidly at first and then more slowly with perhaps a small amount of residual extension remaining. This remaining extension is known as permanent set. As a simplified explanation the instantaneous extension can be considered to be composed of two quantities, the elastic extension, which is completely recoverable, and the plastic or permanent extension, which is not recoverable. According to this simplified

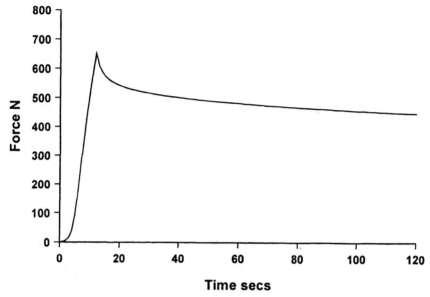

5.14   The decay of load with time.

theory when a material is subject to forces below its yield point then most of the extension is recoverable, whereas if the force is sufficient to take the material beyond its yield point a fraction of the extension will be permanent. This simplified theory does not account for the creep or stress relaxation behaviour which can only be understood if the viscoelastic properties of textile materials are taken into consideration.

A more complex behaviour due to viscoelastic properties can be found when a yarn is cycled between two different force levels [3]. If the yarn is allowed to relax from the higher of the two force levels, the force falls off with time as described above. If, however, the yarn is subjected to a number of cycles between the two forces and then allowed to relax from the lower force level the force increases and reaches a steady value with time, instead of decreasing.

5.3.5   Elastic recovery

When textile materials are stretched by forces that are below the level of their breaking strength and are then allowed to recover, they do not immediately return to their original length. How much of the original length they recover depends on the force used, the length of time the force is applied for and the length of time allowed for the recovery. Farrow [4] uses the following equation as a measure of how much a material recovers its original length after deformation:

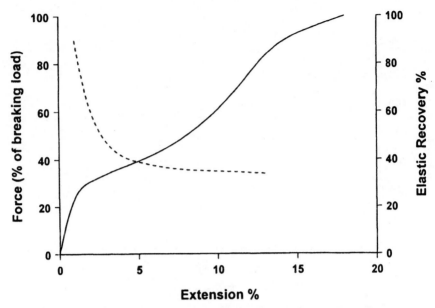

5.15   Dotted line shows the decrease in elastic recovery of acrylic
yarn (RH scale) with increasing extension. Solid line shows the
force extension curve for the yarn, LH scale. From data in [4].

$$\text{Elastic recovery} = \frac{\text{recovered extension}}{\text{imposed extension}} \times 100\%$$

Perfectly elastic materials will have a value of 100%. It was found experimentally [4] that the percentage recovery decreased steadily with increasing extension of the material up to the yield point where the recovery
decreased sharply. This is shown in Fig. 5.15 where a plot of elastic recovery against extension is superimposed on the force extension curve for
acrylic fibre. The elastic recovery changes in nature in the region of the yield
point.

The elastic recovery is also a time-dependent phenomenon, the recovery
depending, among other factors, on the time the material is held at a given
extension [5, 6]. The longer it is held at a given extension, the lower is the
level of recovery. This effect is shown in Fig. 5.16 where the elastic recovery of a sample of cellulose acetate from a fixed extension of 5% is plotted
against the time held at that extension. The recovery is comparatively large
for very short periods of time under extension but decreases quite markedly
when held at that extension for longer periods. If, on the other hand, the
material is held at a given extension for a fixed length of time before
removal of the force, the elastic recovery increases with time, rapidly at first
and more slowly later [5, 6] as shown in Fig. 5.17.

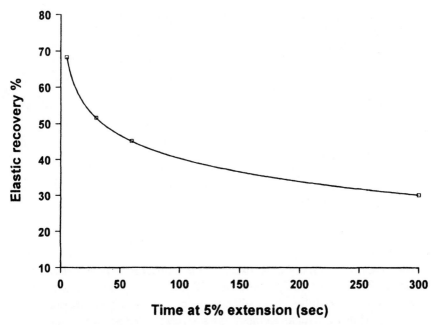

5.16 Decrease in elastic recovery of cellulose acetate with increased time held at 5% extension. From data in [5].

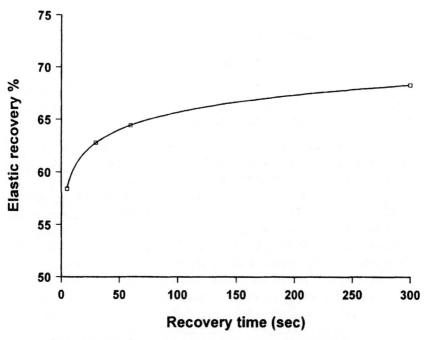

5.17 Increase in elastic recovery of cellulose acetate with recovery time from 5% extension. From data in [5].

Table 5.1 Elastic recovery of carpet fibres from 59% extension (5 min extension, 5 min recovery time in each case)

Fibre type	Recovery (%)
Evlan	34
Fibro	40
Courtelle	54
Nylon	85
Polypropylene	85
Wool	72

Source: from [3].

The extent of recovery from extension is a property which is dependent on the type of material [6] as shown in Table 5.1. This difference helps to account for the variation in resilience properties which are displayed by these materials in diverse applications such as the resistance to flattening of carpet tufts, the recovery of fabrics from creasing and the resistance of fabrics to abrasion.

5.4   Factors affecting tensile testing

5.4.1 Type of testing machine

There are three different ways of carrying out tensile tests with regard to the way of extending the specimen, each of which is historically associated with a particular design of testing instrument:

1 Constant rate of extension (CRE) in which the rate of increase of specimen length is uniform with time and the load measuring mechanism moves a negligible distance with increasing load.
2 Constant rate of traverse (CRT) in which the pulling clamp moves at a uniform rate and the load is applied through the other clamp which moves appreciably to actuate a load measuring mechanism so that the rate of increase of load or elongation is usually not constant and is dependent on the extension characteristics of the specimen. This type of mechanism is usually associated with older types of machine where the load is applied by swinging a weighted pendulum through an arc. The angle that the pendulum has travelled through at the breaking point is then a measure of load. The mechanism is arranged to record the maximum height of the pendulum.

3 Constant rate of loading (CRL) in which the rate of increase of the load is uniform with time and the specimen is free to elongate, this elongation being dependent on the extension characteristics of the specimen at any applied load.

Most modern machines operate on the constant rate of extension principle where the moving jaw is driven by a screw thread moving at a constant rotational speed. The construction of the machine depends on its ultimate load capacity. Larger machines have the beam carrying the load cell supported by a separate screw at each end. Some of the smaller models, intended for low load applications, use only one screw, the upper specimen clamp being supported on the end of a cantilever. The most important consideration is that any flexure of the machine, at the maximum load, should be less than the expected accuracy of extension measurement. The extension, in the absence of an extensometer, is derived from measuring the load at fixed time intervals, thus relying on the accuracy of the crosshead speed for deriving the distance travelled. If accurate measurement of extension is required, an extensometer should be used. This piece of equipment monitors the distance apart of two points on the actual specimen, so avoiding any problems of jaw slip or different extension behaviour near the jaws. These accessories are more important for materials with low extensions and are not normally used for most textile applications where strength measurement is the main concern.

The speeds of crosshead movement found on these instruments range from 0.5 to 500 mm/min or up to 1000 mm/min in some cases. These are all relatively slow speeds compared with those encountered in shock loading applications. To achieve higher speeds a completely different form of drive is required such as is found in pendulum testers.

The load in these strength testers is measured via a load cell in which the deflection of a comparatively stiff beam is measured using either a strain gauge or a linear displacement transducer. This gives a system in which the change in position with increasing load is negligible. The accuracy of the load measurement depends on the capacity of the load cell. Most instruments are quoted as being accurate to within ±1% of the indicated load. This accuracy, however, does not extend to the lower end of the load cell range. The Instron 1011 model, for instance, specifies an accuracy of ±1% of the reading or ±0.2% of the load transducer range in use, whichever is the greater. Therefore with a 5000 N load cell with its lowest load range of 500 N this translates to an accuracy of not better than ±1 N. To obtain the greatest accuracy it is necessary to use load cells at the upper end of their capacity limit. This implies that if fibres, yarns and fabrics are all to be tested with the same machine, then three different load cells of appropriate ranges are needed.

5.4.2 Specimen length

The length of sample under test is known as the gauge length and in most textile tests it is equal to the distance between the inner edges of the clamps. This length has an important effect on the measured strength of the material because of the influence of weak spots on the point of failure.

A material when put under stress will always break at its weakest point. Therefore the longer the length of material that is stressed, the greater will be the probability of finding a weak spot within the test length. The value of strength measured will then be that of the weak spot and not an average value for the whole length.

Consider a uniform material of strength 10N but with weak spots of strength 8N every 100mm along its length. If the portion tested in each test is only 10mm long then the probability of a weak spot in that length is 1 in 10. Ten tests will yield nine of 10N and one of 8N which gives an average value of 9.8N. If the test length is increased to 50mm then the probability of a weak spot in the test is increased to five out of ten so that ten tests will give five of 10N and five of 8N, thus giving an average strength of 9.0N. If the test length is 100mm and above then each test will contain a weak spot so that the average strength will be 8.0N.

This is a much simplified illustration: in practice the faults will be randomly distributed along the length of the material with a normal distribution of strengths around the mean value. Figure 5.18 [7] shows the effect of test length on yarn strength for yarns of different coefficients of variation of breaking load. The lines are based on a test length of 200mm as standard. The strength of a fault-free yarn would not change with test length as shown by the line labelled CV = 0%. As the yarn variability increases, the effect of test length on the measured strength also increases as shown by the lines for coefficients of variation of 10% and 20%.

5.4.3 Rate of loading and time to break

The measured breaking load and extension of textile materials is influenced by the rate of extension that is used in the test. The rates of extension that can be used are governed by the maximum speed attainable by the strength tester used. Most universal strength testers have a restricted range of speeds, whereas automatic yarn strength testers can operate at much higher speeds because of the number of tests that are carried out on yarns.

Most materials show an increase in breaking strength with increasing rate of extension together with a decrease in extension [7]. However, some materials [8–10] reach a maximum at speeds below the highest tested and then show a slight fall. The changes in strength at increasing rates of extension are due to the more or less viscoelastic nature of textile materials which

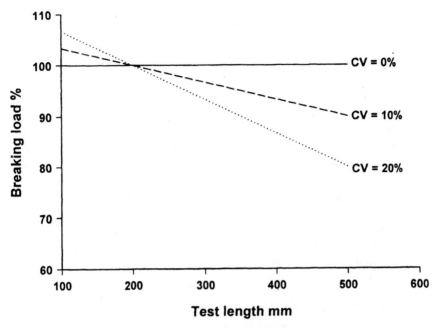

5.18   The effect of testing length on breaking force.

means that they require a certain time to respond to the applied stress. Different types of fibre respond differently and also different yarn and fabric constructions react differently. With filament yarns the stress is directly applied to the fibres so their response depends on the fibre type, whereas in the case of staple fibre yarns there has to be a realigning of the individual fibres in order to spread the load. Vangheluwe [9] shows that for cotton yarns the modulus increases with increasing strain rate although the tenacity and extension reach maxima at intermediate speeds.

5.4.4   Effect of humidity and temperature

Humidity of the testing atmosphere greatly affects the strength and extension of textile materials. This is to assume that the material is in equilibrium with the testing atmosphere as it is the water content of the fibres that matters. The effect varies with the regain of the fibre; hydrophobic materials are hardly affected whereas those with high regains change the most. Wool, silk and viscose lose strength and cotton, linen and bast fibres increase in strength. The difference in the load extension curve between wet and dry wool is shown in Fig. 2.1.

The temperature of the test does not have such a large effect within the range of normal room temperatures at which tests may be carried out. At very low temperatures some fibres may become brittle and at higher temperatures fibre strength may be degraded.

5.4.5 Previous history of the specimen

Changes that a material has undergone prior to it being tested may have a large effect on the measured values of strength and elongation. For example, a specimen may have been strained beyond its yield point in which case its measured strength and elongation will be different from the original material. Alternatively stretching an undrawn or partially oriented material will increase its draw ratio and so increase its strength. A material that has had some form of chemical treatment such as bleaching or that has been exposed to light may be degraded by such treatment and so have lower properties than the original material. Indeed, tensile tests may be used in fault finding to determine whether the material has been overstretched or been subject to chemical degradation.

5.4.6 Clamping problems

During a tensile test textile materials are normally clamped between the faces of two jaws by lateral pressure. This clamping arrangement can give rise to two sorts of problem: slippage of the sample at the jaws or damage of the sample by the jaws, depending on whether the clamping pressure used is too low or too high.

Jaw slip

The total clamping force holding the sample in place is governed by the friction of the clamp faces, the clamping pressure and the length of the jaw in contact with the specimen. If the clamping pressure is low, part of the specimen within the jaws can extend as well as that part of the specimen outside the jaws that is being strained. In Fig. 5.19(a) point A, which was initially at the edge of the clamp before the test began, has moved out of the clamp by a small amount as shown in Fig. 5.19(b) just before failure. This means that the measured extension is higher than the real value. The problem has a greater effect at low clamping pressures and with low friction jaw faces. The amount of jaw slip can be estimated by measuring the elongation at various test lengths and extrapolating the resulting graph of elongation against test length to zero test length where there should be zero elongation. Any elongation above zero at this point is then due to slippage.

A related problem is that of the whole sample slipping through the jaws eventually pulling out in some cases. The problem may only be detected

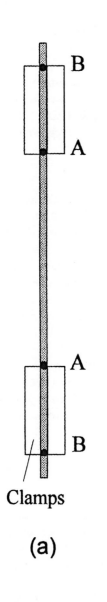

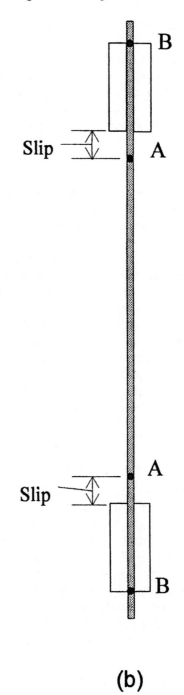

5.19   Jaw slip.

by observing the specimen for movement during the test. If the movement is undetected the recorded extension will be higher than the actual value. Alternatively only part of the yarn or fabric may slip through the jaws due to the clamping of the outer edges of the sample not being as efficient as that of the main part. This problem lowers the recorded strength as not all the elements of the sample are contributing to the strength. Clamp jaws that are soft enough to mould to the specimen can help with this problem.

Jaw breaks

Jaw breaks are premature breaks due to damage to the test specimen by the clamps. They are identified as such because they occur close to the jaw edge and they have the effect of reducing the measured strength of the sample. The problem is particularly acute with hard jaw faces because the clamping forces needed are higher due to their low friction. Jaws with rubber faces which are soft and have a high coefficient of friction reduce the problem and are often used to grip fabrics. Capstan type grips are often used for yarns. With these the yarn is led round a smooth surface of large radius before the actual clamp so that part of the load is taken by the surface friction between the yarn and the capstan.

   High-tenacity yarns and fabrics such as Kevlar and carbon fibre are particularly susceptible to these problems because the gripping forces need to be high. The fibre strength is also easily reduced by any damage due to gripping. In such cases special ones may have to be designed in order to spread the load.

5.5   Fibre strength

Carrying out strength tests on fibres is difficult and time consuming. This is because, particularly with natural fibres, the individual strengths of the fibres vary a great deal and therefore a large number have to be measured to give statistical reliability to the result. Furthermore individual fibres are difficult to handle and grip in the clamps of a strength testing machine, a problem that increases as the fibres become finer. For these reasons single fibre strength tests are more often carried out for research purposes and not as routine industrial quality control tasks. Tests on fibre bundles, which overcome the problems of fibre handling and number of tests needed for accuracy, are carried out as part of the normal range of tests on cotton fibres.

5.5.1 Single fibre strength

Tests on single fibres can be carried out on a universal tensile tester if a suitably sensitive load cell is available. Also required are lightweight clamps

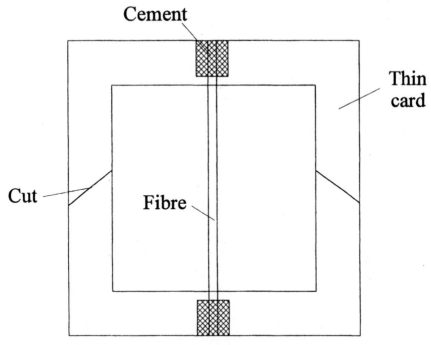

5.20   A holder for single fibres.

that are delicate enough to hold fibres whose diameters may be as low as 10–20 μm. A problem encountered when testing high-strength fibres is that of gripping the fibres tightly enough so that they do not slip without causing jaw breaks due to fibre damage. If the fibres cannot be gripped directly in the testing machine jaws they are often cemented into individual cardboard frames which are themselves then gripped by the jaws. The cardboard frames, shown in Fig. 5.20, have an opening the size of the gauge length required. When they are loaded into the tensile tester the sides of the frame are cut away leaving the fibre between the jaws. The cement used is responsible for gripping the fibres, therefore the samples have to be left for a sufficient time in the frames for the cement to set.

   In addition to the standard strength testing machines there are a number of instruments available solely for fibre strength testing. These include the WIRA single fibre strength tester, the Lenzing Vibrodyn and the Textechno Fafegraph HR. The advantage of these machines is the easier loading of the fibre specimen due to the special clamping arrangements in use.

   The US standard for single fibre strength [11] specifies a gauge length of either $\frac{1}{2}$ in or 1 in (12.7 or 25.4 mm). Up to 40 fibres should be tested depending on the variability of the results. The elongation rate depends on the expected breaking elongation

under 8%	10% of initial specimen length/min
8–100%	60% of initial specimen length/min
over 100%	240% of initial specimen length/min

With fibres that have crimp a pretension of 0.3–1 gf/tex (2.9 to 9.81 mN/tex) can be used to remove the crimp.

The British standard [12] specifies gauge lengths of 10, 20 or 50 mm with a testing speed adjusted so that the sample breaks in either 20 or 30 s. The number of tests is 50 and the level of pretension is set at 0.5 gf/tex (4.9 mN/tex).

5.5.2 Bundle strength

Pressley fibre bundle tester

The Pressley tester is an instrument for measuring the strength of a bundle of cotton fibres. Before they are mounted in the instrument the cotton fibres are combed parallel using a hand comb into a flat bundle about 6 mm wide. The special leather-faced clamps are removed from the machine and placed in a mounting vice so that they lie adjacent to each other, thus giving zero specimen length. The bundle is placed across the two jaws and clamped in position by the top jaws. When the clamps are removed from the mounting device the fringe of fibres protruding from the outer edges of the clamps is trimmed off leaving a known length of fibre within the jaws.

When the jaws are loaded into the instrument the upper jaw of the pair is linked to the short arm of a pivoted beam. The longer arm of the beam is inclined at a small angle to the horizontal and has a weight on it which can roll down the slope. As the weight moves away from the pivot the force on the top jaw gradually increases until the bundle breaks. When this happens the moving weight is automatically halted so that the distance along the arm can be measured. As the distance from the pivot is proportional to the force on the fibre bundle, the arm can be directly calibrated in units of force (lbf). At the end of the test the two halves of the bundle are weighed, and as the total length of the bundle is fixed a figure of merit known as the Pressley Index can be calculated:

$$PI = \frac{force\ (lbf)}{mass\ (mg)}$$

The result can be expressed as gram force per tex by multiplying the index by 5.36 or in mN per tex by multiplying by 52.58. Because the gauge length in this test is zero the extension of the fibres cannot be measured.

Stelometer

The Stelometer is a bundle testing instrument which is capable of measuring elongation as well as strength. The instrument uses the same type of jaws as the Pressley instrument but they have a separation of 3.2 mm ($\frac{1}{8}$ in) as distinct from the zero separation of the Pressley instrument.

The loading of the specimen is carried out by a pendulum system which is mounted in such a way that it rotates about its centre of gravity. This eliminates any inertial effects in loading of the sample which is generally a problem with systems that apply the force using a pendulum. The layout of the instrument is shown in Fig. 5.21: the pendulum is pivoted from the beam but the pivot of the beam is at the centre of gravity of the pendulum. The sample is held between the clamp attached to the beam and the one attached to the pendulum. The beam and the pendulum start in a vertical position but the centre of gravity of the beam is such that when it is released

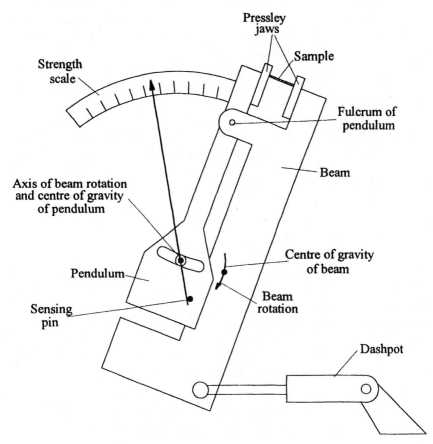

5.21   The Stelometer.

at the start of the test the whole assembly rotates. As the beam rotates the pendulum moves from the vertical so that it then exerts a force on the sample. The speed of rotation of the beam is altered by dashpot so that the rate of loading is 1 kgf/s. A pointer is moved along a scale graduated in breaking force by the sensing pin on the pendulum. When the sample breaks the pendulum falls away leaving a maximum reading. A separate pointer, not shown in the diagram, indicates the sample extension.

After breaking the bundle all the fibres are weighed allowing the tenacity to be calculated:

$$\text{Tenacity in gf/tex} = \frac{\text{breaking force in kgf} \times 15}{\text{sample mass in mg}}$$

The effective total length of the sample is 15 mm (0.590 in) for a $\frac{1}{8}$ in (3.2 mm) gauge length and 11.81 mm (0.465 in) for a zero gauge length so that 11.8 should be used in the above formula if a zero gauge length is used.

The tenacity measured at zero length is greater than that measured at $\frac{1}{8}$ in length because of the general effect that shorter gauge lengths have on measured strength. The ratio between the two values will vary with the variability of the material being tested.

The bundle strength of cotton fibres is also measured as part of the high-volume instrument (HVI) set of tests marketed by Motion Control, Inc. and Special Instruments Laboratory, Inc. As part of these tests a fibre beard is formed whose mass is measured at a number of points along the fibre length to form a fibrogram. Based on the results from the fibrogram a point is selected at a certain distance from the clamp to perform a strength test using jaws with a $\frac{1}{8}$ in (3.2 mm) separation. Taylor [13] has compared the results from these instruments with those from the standard tenacity tests.

5.6   Yarn strength

The strength and extension results from samples of yarn taken from different parts of a package can be very variable. Yarn made from staple fibres is worse in this respect than yarn made from continuous filaments owing to the fact that the number of fibres in the cross-section of a staple fibre yarn is variable. This means that in order to get a reasonable estimation of the mean strength of a yarn a large number of tests have to be carried out on it. Two types of yarn test are carried out:

1   Tests on single lengths of yarn, usually from adjacent parts of the yarn package. These are sometimes referred to as single thread tests.
2   Tests on hanks or skeins of yarn containing up to 120 metres of yarn at a time which is broken as one item.

5.6.1   Yarn strength: single strand method

Most yarn test standards are very similar. The British Standard [14] lays down that the number of tests should be:

1   Single yarns
   (a) continuous-filament yarns: 20 tests,
   (b) spun yarns: 50 tests.
2   Plied and cabled yarns: 20 tests.

The yarns should be conditioned before testing in the standard atmosphere. The testing machine is set to give a test length of 500 mm and the speed is adjusted so that yarn break is reached in 20 ± 3 s. Before each test a pre-tension of 0.5 cN/tex is applied to the yarn in order to give a reproducible extension value. The mean breaking force, mean extension at break as a percentage of the initial length, CV of breaking force and CV of breaking extension are recorded. The US standard [15] specifies a gauge length of 10 ± 0.1 in (250 ± 3 mm) or alternatively by agreement 20 ± 0.2 in (500 ± 5 mm) and uses a time to break of 20 ± 3 s.

Because of the large number of results needed for yarn testing, automatic strength testers are available which will carry out any number of tests on a number of different packages without any operator attention. Uster which produces the Tensorapid automatic strength tester, has compiled a booklet of statistics [37] of yarn strengths of various compositions, spinning routes and linear densities. The intention is that individual test results can be compared with the appropriate statistics to see whether the strength falls into the expected range of values. In these statistics breaking strengths are also given for a high rate of extension, that is 5000 mm/min.

In many uses it is not the mean strength of the yarn that is important but the frequency of any weak places. These lead to the yarn breaking during weaving for example and so give rise to machine stoppages or faults in the fabric that must be avoided for profitable production. Weak places may be hundreds of metres apart but still cause problems in high-speed production. Therefore it is the variability of the yarn tensile properties as measured by the coefficient of variation of the strength and extension that is of greater importance than the mean values in such cases. The aim of yarn quality control is, by the use of statistics, to predict the infrequent occurrence of weak spots. The trend is to test a greater total length of yarn using higher speeds because otherwise the tests would take too long if the standard test time of 20 s was used. More tests will enable better prediction of the statistically few weak spots as 50 tests may only test the first 50 m of a yarn package. However, a balance has to be struck between making too few tests and wasting a large percentage of the yarn package.

5.6.2 Yarn strength: skein method

In this method a long length of yarn is wound into a hank or skein using a wrap reel as would be used for linear density measurement, the two loose ends being tied together. The whole hank is then mounted in a strength testing machine between two smooth capstans, which may be free to rotate. The hank is subjected to increasing extension while the force is monitored. When one part of the yarn breaks, the hank begins to unravel. If the yarn was looped over frictionless pulleys, once one end broke the yarn would then unwrap completely and the strength per strand that was measured would be that of the weakest spot. Because of the friction present in the system the force continues to increase until sufficient strands have broken for the hank to unravel, the force passing through a maximum value at some point. This maximum force is known as the hank strength. Because the friction of the yarn against the pulleys plays a large part in the result, the measured hank strength can vary according to yarn friction and the particular machine that it is measured on.

Measuring the strength of a hank or skein of yarn is a method that was used in the early days of textile testing but that is now being replaced by the single strand method, especially since the development of automatic strength testing machines. The main advantage of the hank method is that it tests a long length of yarn in one test. The yarn is expected to break at the weak spots so giving a more realistic strength value and also the same hank can be used for measuring the yarn count. The disadvantage of the test is that it is dependent on the friction between the yarn and the capstans which determines how well the load is spread between the multiple strands of the hank. This means that the results are specific to a particular machine and yarn combination. The test is considered satisfactory for acceptance testing of commercial shipments but not for measurements which have to be reproducible between laboratories. There is a correlation between the tenacity of yarn measured by the skein method and that measured by the single strand method. The value for the skein is always lower than that for the single strand [10]. Other drawbacks to the method are that there is no measure of strength variability and no measure of yarn extension as the distance moved by the capstans is determined by yarn extension and hank unravelling.

The British Standard [16] specifies a hank of 100 wraps of 1 m diameter. This is tested at such a speed that it breaks within 20 ± 3 s, or alternatively a constant speed of 300 mm/min is allowed. If the yarn is spun on the cotton or worsted systems 10 skeins should be tested and 20 skeins if the yarn is spun on the woollen system. The method is not used for continuous filament yarns.

The US standard [17] has three options for hank size:

1 Eighty, 40 or 20 turns on a 1.5m (1.5yd) reel tested at a speed of 300mm/min. Twenty or 40 turns are to be used when the machine capacity is not great enough to break a hank of 80 turns.
2 Fifty turns on a 1.0m (1yd) reel broken at a speed of 300mm/min.
3 Fifty turns on a 1.0m (1yd) reel broken in a time of 20s.

The number of samples tested is the same as that for the British Standard.

The breaking force per strand increases slightly as the perimeter of the skein is reduced as would be expected from a change in gauge length. The breaking strength of a 1yd skein is 5% higher than that for a 1.5yd skein.

Count strength product

The count strength product (CSP or LCSP) is a measure used for cotton yarns and is the product of the yarn count and the lea (hank) strength. It is based on measuring the strength of an 80 turn hank made on a 1.5yd wrap reel to give a total length of 120yd. The strength is usually measured in pounds force (lbf). The value enables a comparison to be made among yarns of a similar but not necessarily identical count in the same way that tenacity values are used.

Assuming that all 160 strands of the hank have the same strength as the single yarn, the tenacity can be related to the count strength product by the following formula:

$$\text{Tenacity (N/tex)} = 0.000,047 \times \text{CSP (lbf} \times \text{CC)}$$

where CC = cotton count value

5.7   Fabric strength

5.7.1 Strip strength

The British Standard [18] for fabric tensile strength involves extending a strip of fabric to its breaking point by a suitable mechanical means (Fig. 5.22) which can record the breaking load and extension. Five fabric samples are extended in a direction parallel to the warp and five parallel to the weft, no two samples to contain the same longitudinal threads. The specimens are cut to a size of 60mm × 300mm and then frayed down in the width equally at both sides to give samples which are exactly 50mm wide. This ensures that all the threads run the full length of the sample so contributing to the strength and also that the width is accurate. The rate of extension is set to 50mm/min and the distance between the jaws (gauge length) is set to 200mm. The sample is pretensioned to 1% of the probable breaking load. Any breaks that occur within 5mm of the jaws should be rejected and also those at loads substantially less than the average.

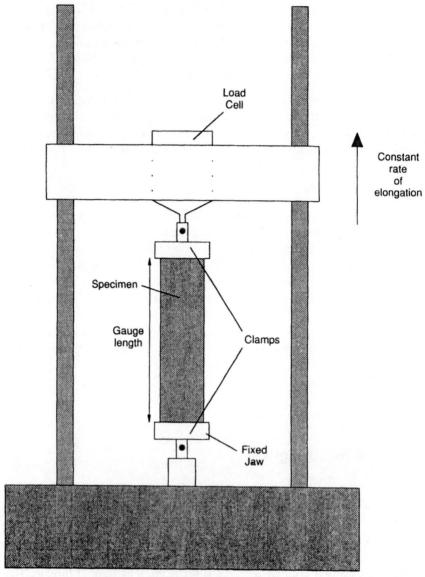

5.22 The apparatus for a fabric tensile test.

The mean breaking force and mean extension as a percentage of initial length are reported.

5.7.2 Grab test

The US Standard [19] contains three ways of preparing the fabric specimen for tensile testing. They are: (1) ravelled strip in 1 in (25 mm) and 2 in

(50 mm) widths where the method of preparation is the same as in the above standard; (2) cut strip in 1 in (25 mm) and 2 in (50 mm) widths which is intended to be used with fabrics such as heavily milled ones which cannot easily be frayed and (3) the grab method which is fundamentally different from the above two methods. The grab test uses jaw faces which are considerably narrower than the fabric, so avoiding the need to fray the fabric to width and hence making it a simpler and quicker test to carry out. The sample used is 4 in (100 mm) wide by 6 in (150 mm) long but the jaws which are used have one of their faces only 1 in (25 mm) wide. This means that only the central 25 mm of the fabric is stressed. A line is drawn on the fabric sample 1.5 in (37 mm) from the edge to assist in clamping it so that the same set of threads are clamped in both jaws. The gauge length used is 3 in (75 mm) and the speed is adjusted so that the sample is broken in 20 ± 3 s. The mounting of the sample in the jaws is shown in Fig. 5.23. In this test there is a certain amount of assistance from yarns adjacent to the central stressed area so that the strength measured is higher than for a 25 mm ravelled strip test.

5.8   Tear tests

A fabric tears when it is snagged by a sharp object and the immediate small puncture is converted into a long rip by what may be a very small extra effort. It is probably the most common type of strength failure of fabrics in use. It is particularly important in industrial fabrics that are exposed to rough handling in use such as tents and sacks and also those where propagation of a tear would be catastrophic such as parachutes. Outdoor clothing, overalls and uniforms are types of clothing where tearing strength is of importance.

5.8.1 Measuring tearing strength

The fabric property usually measured is the force required to propagate an existing tear and not the force required to initiate a tear as this usually requires a cutting of threads. As part of the preparation of the fabric specimens a cut is made in them and then the force required to extend the cut is measured. This is conveniently carried out by gripping the two halves of the cut in a standard tensile tester. The various tear tests carried out in this manner differ mainly in the geometry of the specimen. The simplest is the rip test where a cut is made down the centre of a strip of fabric and the two tails pulled apart by a tensile tester. The test is sometimes referred to as the single rip test, the trouser tear or in the US as the tongue tear test Fig. 5.24(a). What is understood in the UK as the tongue tear test has the specimen cut into three tails Fig. 5.24(b) and (c), the central one is gripped in

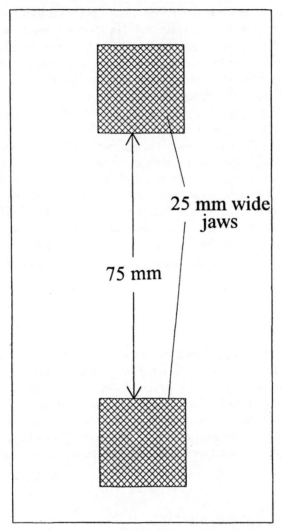

5.23 Grab test sample.

one jaw of the tensile tester and the outer two in the other jaw. This test is also known as the double rip as two tears are made simultaneously.

5.8.2 Single rip tear test

In the US Standard [20] 10 specimens are tested from both fabric directions each measuring 75 mm × 200 mm (3 × 8 in) with an 80 mm (3.5 in) slit part way down the centre of each strip as shown in Fig. 5.24(a). One of the 'tails' is clamped in the lower jaw of a tensile tester and the other side is clamped

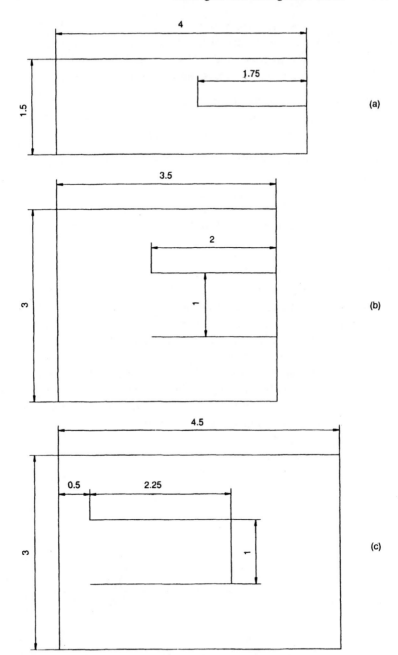

5.24   Tear test samples. All dimensions are in inches.

in the upper jaw, the separation of the jaws causes the tear to proceed through the uncut part of the fabric. The extension speed is set to 50 mm/min (2 in/min) or an optional speed of 300 mm/min can be used.

There are three ways of expressing the result:

1   The average of the five highest peaks.
2   The median peak height.
3   The average force by use of an integrator.

Depending on the direction the fabric is torn in the test is for the tearing strength of filling yarns or of warp yarns.

If the direction to be torn is much stronger than the other direction, failure will occur by tearing across the tail so that it is not always possible to obtain both warp and weft results.

5.8.3 Wing rip tear test

The wing rip test overcomes some of the problems which are found with the single rip test as it is capable of testing most types of fabric without causing a transfer of tear [21]. During the test the point of tearing remains substantially in line with the centre of the grips. The design of the sample is also less susceptible to the withdrawal of threads from the specimen during tearing than is the case with the ordinary rip test. The British standard [22] uses a sample shaped as in Fig. 5.25 which is clamped in the tensile tester in the way shown in Fig. 5.26. The centre line of the specimen has a cut 150 mm long and a mark is made 25 mm from the end of the specimen to show the end of the tear.

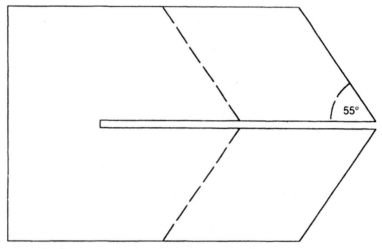

5.25   Wing rip sample layout.

5.26   Sample for tearing strength in tensile tester.

The test is preferred to the tongue tear test though it is not suitable for loosely constructed fabrics which would fail by slippage of the yarns rather than by the rupture of threads.

Five specimens across the weft and five specimens across the warp are tested. The test is carried out using a constant rate of extension testing machine with the speed set at 100 mm/min. The tearing resistance is specified as either across warp or across weft according to which set of yarns are broken.

The results can be expressed as either the maximum tearing resistance or the median tearing resistance. The median value is determined from a force elongation curve such as that shown in Fig. 5.27 and it is the value such that exactly half of the peaks have higher values and half of them have lower values than it. The median tearing resistance value is close to the mean value but it is an easier value to measure by hand methods as it can be determined by sliding a transparent rule down the chart until half the peaks are above the edge of the rule and half below it, at which point the load can be read from the chart.

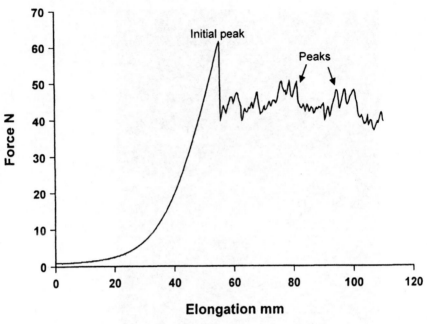

5.27   Tear test force extension curve.

5.8.4 Elmendorf tear tester

The Elmendorf tear tester [23] is a pendulum type ballistic tester which measures energy loss during tearing. The tearing force is related to the energy loss by the following equation:

Energy loss = tearing force × distance
Loss in potential energy = work done

The apparatus which is shown in Fig. 5.28 consists of a sector-shaped pendulum carrying a clamp which is in alignment with a fixed clamp when the pendulum is in the raised starting position, where it has maximum potential energy. The specimen is fastened between the two clamps and the tear is started by a slit cut in the specimen between the clamps. The pendulum is then released and the specimen is torn as the moving jaw moves away from the fixed one. The pendulum possesses potential energy because of its starting height. Some of the energy is lost in tearing through the fabric so that as the pendulum swings through its lowest position it is not able to swing to the same height as it started from. The difference between starting height and finishing height is proportional to the energy lost in tearing the fabric. The scale attached to the pendulum can be graduated to read the tearing force directly or it may give percentage of the original potential energy.

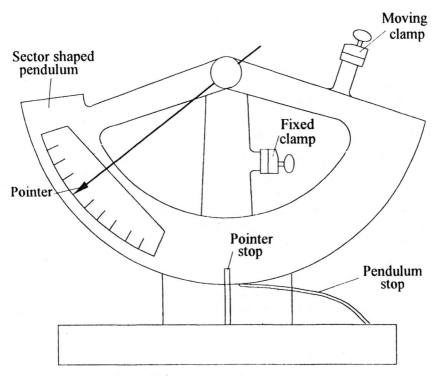

5.28   The Elmendorf tear test.

The apparatus tears right through the specimen. The work done and hence the reading obtained is directly proportional to the length of material torn. Therefore the accuracy of the instrument depends on very careful cutting of the specimen which is normally done with a die. The range of the instrument is from 320 gf to 3840 gf in three separate ranges obtained by using supplementary weights to increase the mass of the pendulum.

When a fabric is being torn all the force is concentrated on a few threads at the point of propagation of the tear. This is why the forces involved in tearing are so much lower than those needed to cause tensile failure. Depending on the fabric construction, threads can group together by lateral movement during tearing, so improving the tearing resistance as more than one thread has to be broken at a time. The peaks that are seen on the load extension curve (Fig. 5.27) are more often from the breaking of a group of threads than from the individual ones. The bunching of threads is also helped by the ease with which yarns can pull out lengthwise from the fabric. The ability to group is a function of the looseness of the yarns in the fabric. Weave has an important effect on this: a twill or a 2/2 matt weave allows the threads to group better thus giving better tearing resistance than a plain weave. High sett fabrics inhibit thread movement and so reduce the assistance effect. Resin treatments such as crease resistance finishes which cause the yarns to adhere to one another also have the same effect. The tensile properties of the constituent fibres have an influence on tearing resistance as those with a high extension allow the load to be shared whereas fibres with low extension such as cotton tear easily. Scelzo *et al.* [24] give a comprehensive account of the factors affecting tear resistance.

5.9   Bursting strength

Tensile strength tests are generally used for woven fabrics where there are definite warp and weft directions in which the strength can be measured. However, certain fabrics such as knitted materials, lace or non-wovens do not have such distinct directions where the strength is at a maximum. Bursting strength is an alternative method of measuring strength in which the material is stressed in all directions at the same time and is therefore more suitable for such materials. There are also fabrics which are simultaneously stressed in all directions during service, such as parachute fabrics, filters, sacks and nets, where it may be important to stress them in a realistic manner. A fabric is more likely to fail by bursting in service than it is to break by a straight tensile fracture as this is the type of stress that is present at the elbows and knees of clothing.

When a fabric fails during a bursting strength test it does so across the direction which has the lowest breaking extension. This is because when stressed in this way all the directions in the fabric undergo the same extension so that the fabric direction with the lowest extension at break is the one that will fail first. This is not necessarily the direction with the lowest strength.

The standard type of bursting strength test uses an elastic diaphragm to load the fabric, the pressure of the fluid behind the diaphragm being used as the measure of stress in the fabric. The general layout of such an instrument is shown in Fig. 5.29. The bursting strength is then measured in units of pressure. As there is a sizeable force needed just to inflate the diaphragm this has to be allowed for in the test. The usual way is to measure the increase in height of the diaphragm during the test and then to inflate the diaphragm to the same height without a specimen present. The pressure required to inflate the diaphragm alone is then deducted from the pressure measured at the point of failure of the sample. The relationship between the diaphragm height and the fabric extension is quite complex so that the method is not used to obtain an estimation of fabric extension.

5.9.1 Diaphragm bursting test

The British Standard [25] describes a test in which the fabric to be tested is clamped over a rubber diaphragm by means of an annular clamping ring

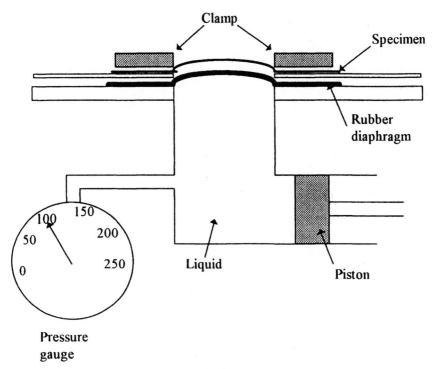

5.29   A diaphragm bursting strength tester.

and an increasing fluid pressure is applied to the underside of the diaphragm until the specimen bursts. The operating fluid may be a liquid or a
gas.

Two sizes of specimen are in use, the area of the specimen under stress
being either 30 mm diameter or 113 mm in diameter. The specimens with
the larger diameter fail at lower pressures (approximately one-fifth of the
30 mm diameter value). However, there is no direct comparison of the
results obtained from the different sizes. The standard requires ten specimens to be tested.

In the test the fabric sample is clamped over the rubber diaphragm and
the pressure in the fluid increased at such a rate that the specimen bursts
within 20 ± 3 s. The extension of the diaphragm is recorded and another
test is carried out without a specimen present. The pressure to do this is
noted and then deducted from the earlier reading.

The following measurements are reported:

Mean bursting strength kN/m^2
Mean bursting distension mm

The US Standard [26] is similar using an aperture of 1.22 ± 0.3 in (31 ± 0.75 mm) the design of equipment being such that the pressure to inflate the diaphragm alone is obtained by removing the specimen after bursting. The test requires ten samples if the variability of the bursting strength is not known.

The disadvantage of the diaphragm type bursting test is the limit to the extension that can be given to the sample owing to the fact that the rubber diaphragm has to stretch to the same amount. Knitted fabrics, for which the method is intended, often have a very high extension.

5.9.2 Ball bursting strength

There is not a British Standard for the ball bursting strength of knitted fabrics although a standard does exist for coated fabrics. This test can be carried out using an attachment on a standard tensile testing machine. In the test a 25 mm diameter steel ball is pushed through the stretched fabric and the force required to do so is recorded. The results are not directly compatible with those from the diaphragm type of bursting tests as they are measured in units of force only and not in units of force per unit area. The advantage of the test is that it can be carried out on a standard universal strength tester with a suitable attachment. There is also no limit to the amount a sample can be extended as there is with the diaphragm test.

The US Standard [27] specifies a 1.0000 in diameter ball (25.4 mm) with a clamp diameter of 1.75 in (44.45 mm) and a speed of 12 in/min (305 mm/min). The standard shows an attachment which is used in the tensile mode on a standard strength testing machine. The British Standard for coated fabrics [28] specifies very similar dimensions with a ball diameter of 25.2 mm, a clamp diameter of 45 mm and a testing speed of 5 mm/s. It is simpler when carrying out this test to use an attachment which operates in the compression mode, if the testing machine is capable of this. An example of a compression fixture to carry out this test is shown in Fig. 5.30.

5.10   Stretch and recovery properties

Certain types of clothing, particularly sports wear, is made to be a close fit to the body. The fabric of which such clothing is made has to be able to stretch in order to accommodate firstly the donning and removal of the clothing and secondly any activity that is undertaken while wearing it. So that the garment remains close fitting and does not appear baggy this stretch has to be followed by the complete recovery of the original dimensions. This is usually accomplished by incorporating a small percentage of elastic fibres into the structure. The requirements of a fabric can be gauged from the

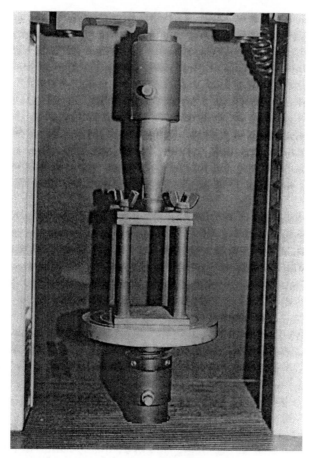

5.30 A ball bursting strength attachment.

typical values of stretch that are encountered during the actions of sitting, bending, or flexing of knees and elbows:

- Back flex 13–16%.
- Elbow flex lengthwise 35–40%, circumferentially 15–22%.
- Seat flex 25–30%, across 6%.
- Knee flex lengthwise 35–45%, circumferentially 12–14%.

Standing at rest is taken as the zero value for the purpose of calculating these increases.

There are a large number of tests devised for stretch fabric by various organisations, all following similar procedures but differing widely in many of the important details. There are two quantities that are generally measured: one is the extension at a given load (sometimes known as modulus) which is a measure of how easily the fabric stretches; the other is how well the fabric recovers from stretching to this load, usually measured as growth or residual extension. As a rule of thumb a stretch fabric is expected to recover to within 3% of its original dimensions. The size of load used in the test is important as a load that is high for a light-weight fabric may not put any serious stretch on a heavy-weight fabric. In general the larger percentage of the breaking load which a fabric is subjected to the greater is the residual extension or growth. The load used is therefore an important test detail and one that differs from test to test. The other details on which the tests differ are the number of stretch cycles before the actual measurements, the time held at the fixed load and the time allowed for recovery.

The British Standard of test for elastic fabrics [29] describes a number of tests for elastic fabrics using either line contact jaws or looped specimens. A gauge length of 100 mm is specified for straight specimens and a total length of 200 mm for looped specimens; the results from the two types of specimens are not necessarily comparable. The standard covers both woven and knitted fabrics. Tests include: extension at a specified force, modulus, residual extension and tension decay.

In the test for extension at a specified force the sample is cycled twice between zero extension and the specified force at a rate of 500 mm/min. The elongation at the specified force is measured on the second cycle from the force extension curve.

The modulus can also be determined from the second cycle by recording the force at specified values of elongation.

In the test for residual extension, gauge length marks are made on the sample which is then clamped in the jaws of a suitable tensile tester so that the marks line up with the inner edge of the clamps. The sample is given one preliminary stretch cycle then extended to a specified force which is held for 10 s. It is removed and allowed to relax on a flat, smooth surface and its length is remeasured after 1 min to see how much of its original length it has recovered to give a length L_2.

If the specification requires it the specimen can be remeasured after 30 min to give a length L_3.

The following quantities are calculated:

$$\text{Mean residual extension after 1 min} = \frac{L_2 - L_1}{L_1} \times 100\%$$

$$\text{Mean residual extension after 30 min} = \frac{L_3 - L_1}{L_1} \times 100\%$$

where L_1 is the original gauge length.

Tension decay is measured by holding the sample at a specified elongation for 5 min and determining the decay in force over this period:

$$\text{Tension decay} = \frac{F_1 - F_2}{F_1} \times 100\%$$

where F_1 = maximum force at specific elongation,
   F_2 = force after 5 min.

The method also covers the fatiguing and ageing of specimens and the subsequent testing any of the above properties.

The US Standard [30] for woven fabrics uses pairs of specimens, one of which is stretched to a fixed load of 4 lb (1800 g) and the other is subsequently held at a fixed extension for 30 min. Three pairs of samples are taken from the stretch direction each 2.5 in × 22 in (64 mm × 560 mm) and are frayed down to 2.0 in width (50 mm). Two lines are marked on the sample 500 mm apart and it is then clamped in the testing machine with the marks aligned with the edges of the jaws. The samples are cycled three times to the fixed load each cycle taking 5 s. At the fourth stretch the load is held for 10 s and the distance between the lines is measured. The load is then removed and the sample remeasured after 30 s. The quantities measured are: percentage stretch and fabric growth, which is the same as residual extension.

If A = original length, B = length under load and C = length after release:

$$\text{Stretch} = \frac{B - A}{A} \times 100\%$$
$$\text{Growth} = \frac{C - A}{A} \times 100\%$$

The second specimen of each pair is stretched to a fixed extension which is taken to be 85% of the stretch measured in the above section or other agreed figure. This extension is held for 30 min, after which the specimen is removed and the growth measured after 30 s and 30 min.

The US Standard [31] covers knitted fabrics having low power. The fabrics are stretched to a fixed extension for 2 h and the growth measured. The amount of extension is governed by the end use of the fabric as shown in Table 5.2.

Five samples are cut from the wale direction and five from the course direction each 5 in × 15 in (127 mm × 398 mm). Each of these is sewn into a loop and bench marks are made 5 in (127 mm) apart on one side. The loops are held by a rod at each end and the appropriate extension from the above table is held for 2 h. After this time the sample is released and the growth measured at 60 s and 1 h after by a rule which is attached to the specimen.

If it is desired the fabric stretch can be measured at a fixed load which is 5 lbf (22.2 N) for the loose fitting fabrics and 10 lbf (44.5 N) for the form

Table 5.2 Percentage stretch for different end uses

	Stretch (%)	
	Course way	Wale way
Loose fitting (comfort stretch)	30	15
Form fitting (semi-support)	60	35

fitting fabrics. The samples are cycled to the given load four times for 5 s before the actual measurement which is taken after the fifth loading has been held for 10 s. The calculations for growth and stretch are the same as in [30] above.

5.11 Seam strength

Failure at a seam makes a garment unusable even though the fabric may be in good condition. There are a number of possible causes of seam failure:

1 The sewing thread either wears out or fails before the fabric does.
2 The yarns making up the fabric are broken or damaged by the needle during sewing.
3 Seam slippage occurs.

Seam slippage is an inherent property of the fabric and so forms part of the specification for fabrics which are to be made into upholstery and apparel. The other problems listed above are specific to making-up and they depend on the sewing machine used, the sewing thread, the sewing speed, size of sewing needle and stitch length among other factors. Similarly seam strength, although it can be measured in the same way as fabric tensile strength, depends on too many factors to be a useful property of a fabric.

5.11.1 Seam slippage

Seam slippage is the condition where a seam sewn in the fabric opens under load. Some of this gap may close on removal of the load but some of it may be a permanent deformation. Seam slippage is a fabric problem especially for fabrics that contain slippery yarns or that have an open structure or where the number of warp and weft interlacings is low. Such factors mean that one set of yarns may be easily pulled through the other. Seam details such as the seam allowance, seam type and stitch rate can all affect the problem but, in tests for seam slippage, these factors are kept constant. The direction of seam slippage tests is often given as weft (filling) over warp when the weft yarns are being pulled through the warp yarns; in this case the seam would be sewn along the weft direction. The opposite direction would be warp over weft.

There are three different types of seam slippage test in existence, each of which has its drawbacks. Firstly there is the type where a standard seam is put under a fixed load and the seam gape is measured. In second type a load extension curve is plotted with and without a standard seam and the difference between the two curves is taken as the slippage. The third type does away with a sewn seam and measures the force required to pull a set of pins through the fabric. A variant of the first type is to measure the load required to give a fixed seam opening.

5.11.2 Seam slippage tests

The British Standard test for seam slippage [32] is a test of the second type. Five warp and five weft specimens each 100 mm × 350 mm are used. Each sample is folded 100 mm from one end and a seam is sewed 20 mm from the fold line using the special sewing thread and sewing machine settings which are detailed in the standard. The layout of the sample is shown in Fig. 5.31. After sewing the folded part of the fabric is cut away 12 mm from the fold line leaving the seam 8 mm from the cut edge. A standard strength tester equipped with 25 mm grab test jaws is used, the gauge length being set to 75 mm.

Just before the test the sample is cut into two parts one with the seam and one without but with each part containing the same set of warp or weft threads. The sample without a seam is first stretched in the tensile tester up to a load of 200 N and a force elongation curve drawn. The matching sample with the seam is then tested in the same way making sure that the force elongation curve starts from the same zero position. The horizontal separation between the curves as shown in Fig. 5.32 is then due to opening of the seam.

In order to find the force required to open the seam a given distance, the separation of the curves at a force of 5 N is measured and this distance is added to the seam opening specified (usually 6 mm but some specifications require 5 mm) making appropriate allowance for the horizontal scale of the chart. Next the point on the curves where there is a separation of this distance is located and the value of load at this point is read off the chart. If the curves do not reach the specified separation below 200 N then the result is recorded as 'more than 200 N'.

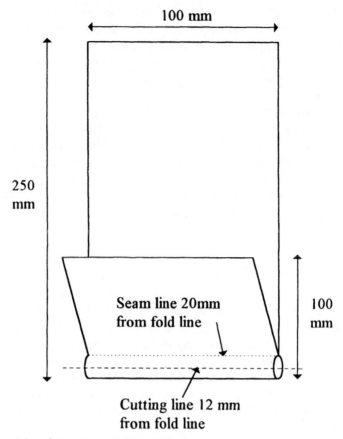

5.31   Seam slippage specimen.

The US Standards [33] and [34] are very similar to the above method except that a load of 1 lbf (4.4 N) is used for correcting for slack in the system instead of 5 N. The required result is the load to produce a seam opening of 6 mm (0.25 in).

5.11.3 Fixed load method

The previous version of BS 3320 used the following method.

Principle

A strip of fabric is folded and stitched across its width. A force is then applied to the strip at right angles to the seam using grab-test jaws and the extent to which the seam opens for a given force is measured.

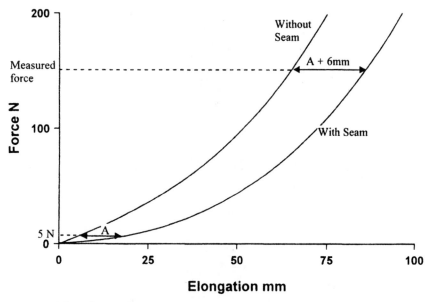

5.32   Seam slippage graph.

The force applied to the seamed specimens is not dictated solely by the mass per unit area of the fabric, instead it depends on the end use of the fabric under test. The most lightly stressed articles such as ladies' dresses, cushions and tickings are tested at a force of 80N; for seams likely to be subjected to greater stress such as in overcoats, suits and overalls a force of 120N is used and for such end uses as upholstery where considerable seam strength is required the seam is stressed to 175N.

Method

Five samples from the warp direction and five samples from the weft direction each 200 mm × 100 mm are used. Each sample is folded in half and a seam is machined 20 mm from the fold, using the special sewing thread and a stitch rate of five stitches per cm. The folded edge is then cut off 12 mm from the fold line. The jaws of the strength tester are set to a gauge length of 75 mm and its speed is set to 50 mm/min. The specimen is clamped in the 25 mm wide jaws so that only the centre portion of the fabric and seam is stressed. The load is increased to either 80, 120 or 175N depending on the end use of the fabric and held at that value for 2 min. The load is then reduced to 2.5N and held there for a further 2 min. The width of the seam opening at its widest place is then measured to the nearest 0.5 mm. The mean value for warpwise and for weftwise specimens is reported.

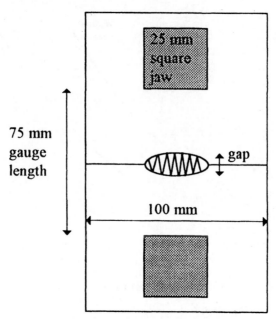

5.33   Specimen for upholstery seam slippage.

5.11.4 Upholstery seam slippage

The British Standard [35] for seams in upholstery is different from the test used for apparel and is of the fixed load type.

Method

Cut five samples from the warp direction and five samples from the weft direction each 200 mm × 100 mm. Fold each sample in half and machine a seam 20 mm from the fold, using the specified sewing thread and stitch details. Cut off the fold 12 mm from the fold line. Set the jaws of the tensile tester 75 mm apart and set the speed to 50 mm/min. Load the specimen in the jaws using the 25 mm jaws so that only the centre of the specimen is clamped as shown in Fig. 5.33. Increase the load to 175 N and hold at that for 2 min. Reduce the load to 2.5 N and hold for a further 2 min. Measure the width of the seam opening as shown in Fig. 5.34, at its widest place to the nearest 0.5 mm. Give the mean value for warpwise and for weftwise specimens.

The difficulty with fixed load methods is that of measuring the seam opening with any accuracy. It is not always obvious where the opening starts and finishes. Methods that use a pair of load extension curves overcome the

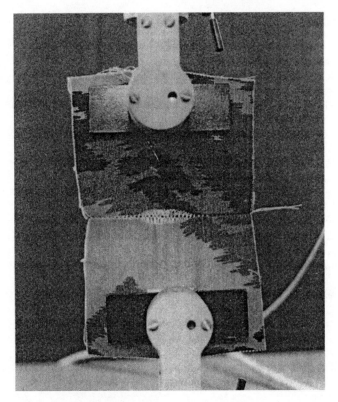

5.34   Seam slippage.

measurement problem but appear to underestimate the amount of slippage due to the high tension allowance (5 N). They also produce the problem of tests with no numerical results due to the seam or fabric failing or the test being stopped before reaching the required seam opening.

The US Standard [36] 'resistance to yarn slippage in woven upholstery fabrics' uses a completely different approach. The fabric sample is impaled on a standard set of pins at a fixed distance from its edge and the maximum force required to pull either the filling yarns over the warp yarns or the warp yarns over the filling yarns is measured. This method avoids the sewing of a standard seam, but the results are not directly comparable with the other methods.

General reading

Morton W E and Hearle J W S, *Physical Properties of Textile Fibres*, 3rd edn. Textile Institute. Manchester. 1993.

References

1. *Textile Terms and Definitions*, 10th edn. McIntyre J E and Daniels P N, eds Textile Institute, Manchester, 1995.
2. Morton W E and Hearle J W S, *Physical Properties of Textile Fibres*, 3rd edn. Textile Institute, Manchester, 1993.
3. Vangheluwe L, 'Relaxation and inverse relaxation of yarns after dynamic loading', *Textile Res J*, 1993 **63** 552.
4. Farrow B, 'Extensometric and elastic properties of textile fibres', *J Text Inst*, 1956 **47** T58.
5. Guthrie J C and Norman S, 'Measurement of the elastic recovery of viscose rayon filaments', *J Text Inst*, 1961 **52** T503.
6. Guthrie J C and Wibberley J, 'The effect of time on the elastic recovery of fibres', *J Text Inst*, 1965 **56** T97.
7. Furter R, *Strength and Elongation Testing of Single and Ply Yarns: Experience with Uster Tensile Testing Installations*, Textile Institute and Zellweger Uster AG, Manchester, 1985.
8. Balasubramanian P and Salhotra K R, 'Effect of strain rate on yarn tenacity', *Text Res J*, 1985 **55** 74.
9. Vangheluwe L, 'Influence of strain rate and yarn number on tensile test results', *Text Res J*, 1992 **62** 586.
10. DeLuca L B and Thibodeaux D P, 'Comparison of yarn tenacity data obtained using the Uster Tensorapid, Dynamat II, and Scott skein testers', *Text Res J*, 1992 **62** 175.
11. ASTM D 3822 Test method for tensile properties of single textile fibres.
12. BS 3411 Method for the determination of the tensile properties of individual textile fibres.
13. Taylor R A, 'Cotton tenacity measurements with high speed instruments', *Tex Res J*, 1986 **56** 92.
14. BS 1932 Testing the strength of yarns and threads from packages.
15. ASTM D 2256 Test method for breaking load (strength) and elongation of yarn by the single strand method.
16. BS 6372 Method for determination of breaking strength of yarn from packages: skein method.
17. ASTM D 1578 Test method for breaking load (strength) of yarn by the skein method.
18. BS 2576 Method for determination of breaking strength and elongation (strip method) of woven fabrics.
19. ASTM D 1682 Test methods for breaking load and elongation of textile fabrics.
20. ASTM D 2261 Test method for tearing strength of woven fabrics by the tongue (single rip) method (constant rate of extension tensile testing machine).
21. Harrison P W, 'The tearing strength of fabrics I. A review of the literature', *J Text Inst*, 1960 **51** T91.
22. BS 4303 Method for the determination of the resistance to tearing of woven fabrics by the wing rip technique.
23. ASTM D 1424 Test method for tear resistance of woven fabrics by falling pendulum (Elmendorf) apparatus.

24. Scelzo W A, Backer S and Boyce M C, 'Mechanistic role of yarn and fabric structure in determining tear resistance of woven cloth Part 1 Understanding tongue tear', *Text Res J*, 1994 **64** 291.
25. BS 4768 Method for the determination of the bursting strength and bursting distension of fabrics.
26. ASTM D 3786 Test method for hydraulic bursting strength of knitted goods and non woven fabrics – diaphragm bursting strength tester method.
27. ASTM D 3787 Test method for bursting strength of knitted goods – constant rate of traverse (CRT), ball burst test.
28. BS 3424 Testing coated fabrics Part 6 Methods 8A and 8B. Methods for determination of bursting strength.
29. BS 4952 Method of test for elastic fabrics.
30. ASTM D 3107 Test method for stretch properties of fabrics woven from stretch yarns.
31. ASTM D 2594 Test methods for stretch properties of knitted fabrics having low power.
32. BS 3320 1988 Method for determination of slippage resistance of yarns in woven fabrics: seam method.
33. ASTM D434 Test method for resistance to slippage of yarns in woven fabrics using a standard seam.
34. ASTM D 4034 Test method for resistance to yarn slippage at the sewn seam in woven upholstery fabrics: plain, tufted, or flocked.
35. BS 2543 Specification for woven and knitted fabrics for upholstery Appendix A.
36. ASTM D 4159 Test method for resistance to yarn slippage in woven upholstery fabrics plain, tufted or flocked using a simulated seam.
37. Ulster News, Bulletin No. 31 (Statistics 82), Zellweger Uster, Switzerland, Dec 1982.

6

Dimensional stability

6.1 Introduction

The dimensional stability of a fabric is a measure of the extent to which it keeps its original dimensions subsequent to its manufacture. It is possible for the dimensions of a fabric to increase but any change is more likely to be a decrease or shrinkage. Shrinkage is a problem that gives rise to a large number of customer complaints. Some fabric faults such as colour loss or pilling can degrade the appearance of a garment but still leave it usable. Other faults such as poor abrasion resistance may appear late in the life of a garment and to some extent their appearance may be anticipated by judging the quality of the fabric. However, dimensional change can appear early on in the life of a garment so making a complaint more likely. A recent survey of manufacturers [1] rated shrinkage as one of the ten leading quality problems regardless of the size of the company.

Fabric shrinkage can cause problems in two main areas, either during garment manufacture or during subsequent laundering by the ultimate customer. At various stages during garment manufacture the fabric is pressed in a steam press such as a Hoffman press where it is subjected to steam for a short period while being held between the upper and lower platens of the press.

Laundering is a more vigorous process than pressing and it usually involves mechanical agitation, hot water and detergent. Tumble drying can also affect the shrinkage as the material is wet at the beginning of the drying process, the material being agitated while heated until it is dry. Dry cleaning involves appropriate solvents and agitation; the solvents are not absorbed by the fibres so they do not swell or affect the properties of the fibres. This reduces some of the problems that occur during wet cleaning processes.

There are a number of different causes of dimensional change, some of which are connected to one another. Most mechanisms only operate with fibre types that absorb moisture, but relaxation shrinkage can affect any fibre type. The following types of dimensional change are generally recognised:

1 **Hygral expansion** is a property of fabrics made from fibres that absorb moisture, in particular fabrics made from wool. It is a reversible change in dimensions which takes place when the moisture regain of a fabric is altered.
2 **Relaxation shrinkage** is the irreversible dimensional change accompanying the release of fibre strains imparted during manufacture which have been set by the combined effects of time, finishing treatments, and physical restraints within the structure.
3 **Swelling shrinkage** results from the swelling and de-swelling of the constituent fibres of a fabric due to the absorption and desorption of water.
4 **Felting shrinkage** results primarily from the frictional properties of the component fibres which cause them to migrate within the structure. This behaviour is normally considered to be significant only for fibres having scales on their surface such as wool.

The dimensions of fabrics can become **set** while they are deformed if they are subjected to a suitable process. Fibres that absorb water can be set if they are deformed while in the wet state and then dried at those dimensions. Thermoplastic fibres can be set if they are deformed at a comparatively high temperature and then allowed to cool in the deformed state. The set may be temporary or permanent depending on the severity of the setting conditions. During relaxation shrinkage it is temporary set that is released. It is generally the case that deformation that has been set can be released by a more severe treatment than the setting treatment. Conversely if it is wished to make the dimensions of the fabric permanent it is necessary to carry out the setting at conditions that the fabric will not meet in use.

Figure 6.1 shows the relative scale of severity for treating wool. Any setting that has taken place under one of the conditions listed can be relaxed by subjecting the material to any of the conditions above it on the list.

6.1.1 Hygral expansion

Hygral expansion refers to the property of certain fabrics that absorb moisture, where the fabric expands as the moisture content increases, owing to the swelling of the constituent fibres. This is particularly a property of wool fabrics. All of the expansion is subsequently reversed when the fabric is dried to its original moisture content. The increase in dimensions takes place in both warp and weft directions and its magnitude is related to the amount of moisture in the material. Figure 6.2 shows the increase in

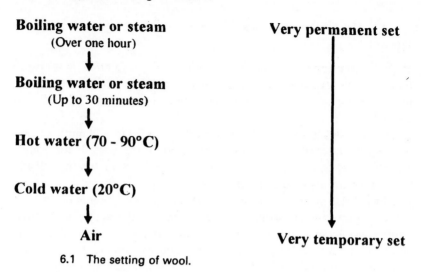

Boiling water or steam
(Over one hour)
↓
Boiling water or steam
(Up to 30 minutes)
↓
Hot water (70 - 90°C)
↓
Cold water (20°C)
↓
Air

Very permanent set

Very temporary set

6.1   The setting of wool.

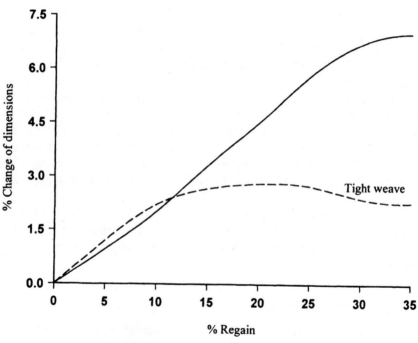

6.2   The hygral expansion of wool.

dimensions of two wool fabrics with increasing atmospheric moisture content; in one case the expansion increases with regain almost up to the maximum value for wool, whereas in the other fabric the expansion reaches a maximum at around 20% regain. This is considered [2, 3] to be due to the tighter weave of the second fabric which causes the widthways expansion of the warp yarn to interfere with the lengthways expansion of the weft yarn. Hygral expansion is believed to be caused [2] by the straightening of crimped yarn as it absorbs moisture. This is due to the fact that wool fibres swell to 16% in diameter and 1% in length when wet. The swelling causes fibres which have been permanently set into a curve to try to straighten out due to the imbalance of forces. When the fibres dry out they revert to their former diameter and so take up their original curvature.

The increase in dimensions due to hygral expansion can take place in the fabric at the same time as any shrinkage because of relaxation of set in stresses such as occurs when the fabric is soaked in water. The magnitude of the expansion can in fact be greater than that of the shrinkage. The dimensions of a fabric when it is first wetted out and then dried depend on its moisture content and are a combination of the increase due to hygral expansion and any decrease due to shrinkage. It is for this reason that in making shrinkage measurements, a fabric should always have the same moisture content during the final measurement as it had when the initial measurement was made.

Hygral expansion of a fabric in a finished garment can cause problems when the garment is exposed to an atmosphere of higher relative humidity than that in which it was made. The expansion can cause pucker at seams and wrinkling where it is constrained by other panels or fixed interlinings.

6.1.2 Relaxation shrinkage

When yarns are woven into fabrics they are subjected to considerable tensions, particularly in the warp direction. In subsequent finishing processes such as tentering or calendering this stretch may be increased and temporarily set in the fabric. The fabric is then in a state of dimensional instability. Subsequently when the fabric is thoroughly wetted it tends to revert to its more stable dimensions which results in the contraction of the yarns. This effect is usually greater in the warp direction than in the weft direction.

Relaxation shrinkage in wool fabrics is caused by stretching the wet fabric beyond its relaxed dimensions during drying. A proportion of the excess dimensions is retained when the dry fabric is freed of constraint. The fabric will, however, revert to its original dimensions when soaked in water. This effect is related to the hygral expansion value of a fabric [3] in that a fabric with a high value of hygral expansion will increase its dimensions more when it is wetted out so that it subsequently needs to contract to a greater extent when it is dried. Merely holding such a fabric at its wet dimensions will thus give rise to a fabric that is liable to relaxation shrinkage.

6.1.3 Swelling shrinkage

This type of shrinkage results from the widthways swelling and contraction of the individual fibres which accompanies their uptake and loss of water. For instance viscose fibres increase in length by about 5% and in diameter by 30–40% when wet [4]. Because of the fibre swelling, the yarns made from them increase in diameter which means that, for instance, a warp thread has to take a longer path around the swollen weft threads. This is shown diagrammatically in Fig. 6.3 where the swelling of the yarns from the dry state (a) to the wet state (b) causes an increase in the length of the path the yarn must take if the fibre centres remain the same. In a fabric the warp yarn must either increase in length or the weft threads must move closer together. In order for the warp yarn to increase in length, tension needs to be applied to the fabric to stretch it. In the absence of any tension, which is usually the case during washing, the weft threads will therefore move closer together. Although the fibre dimensions will revert to their original values on drying, the forces available for returning the fabric to its original dimensions are not as powerful as the swelling forces so that the process tends to be one way. The overall effect of the swelling mechanism on a fabric's dimensions is dependent on the tightness of the weave [4]. This mechanism is the one that is active when viscose and cotton fabrics shrink.

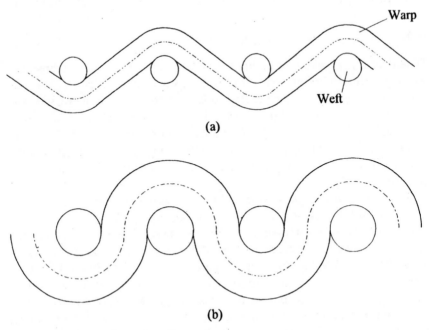

(a)

(b)

6.3 The effect of swelling of yarns.

Processes have been developed to pre-shrink cotton material in finishing to reduce the amount of shrinkage in consumer goods.

6.1.4 Felting shrinkage

Felting shrinkage is a mechanism of shrinkage that is confined to wool fabrics and it is a direct consequence of the presence of scales on the wool surface as shown in Fig. 6.4. Deliberate use of this effect is made in milling to increase the density of structures. Felting is related to the directional frictional effect (DFE) which is found in wool fibres. The coefficient of friction of wool fibres is greater when the movement of the fibre in relation to another surface is in the direction of the tip than when it is in the direction of the root. This effect can be measured directly. Shrinkage is caused by the combined effects of DFE and fibre movement promoted by the elasticity of wool. The behaviour is promoted if the fibres are in warm alkaline or acid liquor.

When alternating compression and relaxation are applied to the wet material, the compression force packs the fibres more tightly together but on relaxation of the force the DFE prevents many of the fibres from reverting to their original positions.

Wool can be made shrink resistant by treatment to reduce the effect of the scales on friction. Chlorine treatments tend to remove the scales;

6.4   The scales on a wool fibre ×1300.

however, too drastic a treatment can reduce the strength of the fibres. Resin treatments are used to mask the scales. The most successful treatments use a combination of the two approaches.

6.1.5 Weft knitted wool fabrics

Knitted fabrics are similar to woven fabrics in that they are subject to relaxation shrinkage and also to felting shrinkage if they are made of wool. However, it has been found difficult experimentally to determine when a fabric has reached a totally relaxed state in which it is in a stable state with the minimum energy. This is because the stable state of a knitted fabric is controlled by the interplay of forces required to shape the interlocking loops of yarn, whereas the stable state of a woven fabric is controlled by the balance of forces required to crimp the yarns. The resistance provided by inter-yarn friction prevents the yarn taking up its lowest energy state and the magnitude of the restoring forces in a knitted fabric is not great enough to overcome this.

Because of this difficulty a number of relaxed states have been suggested [5]:

1 **Dry relaxed state**. This is the condition the fabric reaches after a sufficient period of time subsequent to being removed from the knitting machine.
2 **Wet relaxed state**. This is achieved by a static soak in water and flat drying.
3 **Finished relaxed state**, also known as the consolidated state. This is achieved by soaking in water with agitation, agitation in steam or a static soak at higher temperatures (>90 °C) and drying flat.
4 **Fully relaxed state**. This is achieved by a water soak at 40 °C for 24 h followed by hydro-extraction and tumble drying for 1 h at 70 °C.

The most appropriate treatment will depend on the conditions that the material encounters during finishing.

6.2   Methods of measuring dimensional stability

6.2.1 Marking out samples

The general procedures for preparing and marking out of samples is laid down in the British Standard [6]. Many dimensional stability tests follow very similar lines differentiated only by the treatment given to the fabric, so that these procedures may be followed if no specific test method exists.

For critical work the recommended sample size is 500 mm × 500 mm and for routine work a minimum sample size of 300 mm × 300 mm is consid-

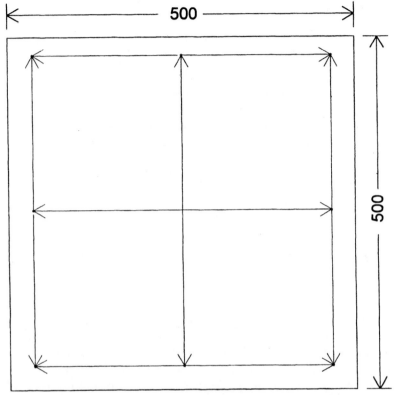

6.5 Marking out of a sample for measuring dimensional stability.

ered sufficient. The samples are marked with three sets of marks in each direction, a minimum of 350 mm apart and at least 50 mm from all edges as shown in Fig. 6.5. In the case of the smaller sample the marks are made 250 mm apart and at a distance of 25 mm from the edge. For critical work it is recommended that the samples are preconditioned at a temperature not greater than 50 °C with a relative humidity of between 10% and 25%. All samples are then conditioned in the standard atmosphere. After measurement the samples are subjected to the required treatment and the procedure for conditioning and measuring repeated to obtain the final dimensions.

6.2.2 WIRA steaming cylinder

The WIRA steaming cylinder [7] is designed to assess the shrinkage that takes place in a commercial garment press as steam pressing is part of the normal garment making up process. The shrinkage that takes place when a

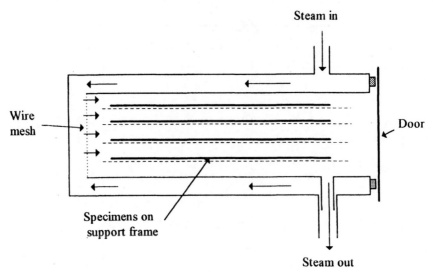

6.6   The WIRA steaming cylinder.

fabric is exposed to steam is classified as relaxation shrinkage not felting or consolidation shrinkage.

In the test the fabric is kept in an unconstrained state and subjected to dry saturated steam at atmospheric pressure. These conditions are slightly different from those that occur in a steam press where the fabric is trapped between the upper and lower platens while it is subjected to steam.

Four warp and four weft samples are tested, each measuring 300 mm × 50 mm. They are first preconditioned and then conditioned for 24 h in the standard testing atmosphere in order that the samples always approach condition from the dry side. Markers (threads, staples, ink dots) are then put on the fabric so as to give two marks 250 mm apart on each sample.

The four specimens are then placed on the wire support frame of the apparatus shown in Fig. 6.6 and steam is allowed to flow through the cylinder for at least one minute to warm it thoroughly. The frame is then inserted into the cylinder keeping the steam valve open and the following cycle carried out:

- Steam for 30 s
- Remove for 30 s

This cycle is performed three times in total with no additional intervals.

The specimens are then allowed to cool, preconditioned and then conditioned for another 24 h to bring them into the same state they were in when they were marked. They are then remeasured on a flat smooth surface and the percentage dimensional change calculated. The mean dimensional change and direction is reported:

$$\text{Shrinkage} = \frac{(\text{original measurement} - \text{final measurement})}{\text{original measurement}} \times 100\%$$

6.2.3 Relaxation shrinkage

The international standard for measuring relaxation shrinkage is the determination of dimensional changes of fabric induced by cold water immersion [8]. In the test the strains in the fabric are released by soaking the fabric without agitation in water that contains a wetting out agent. The specimen is conditioned, measured, soaked in water, dried, reconditioned and measured again.

One specimen of dimensions 500 mm × 500 mm is tested. Three pairs of reference points are made in each direction on the fabric a distance of 350 mm apart and placed not nearer than 35 mm to the edge as shown in Fig. 6.5. When knitted fabrics are to be tested they are folded to give a double thickness with the free edges sewn together.

Before the test the sample is conditioned in the standard atmosphere for 24 h and then laid on a smooth glass surface and covered with another piece of glass to hold it flat while it is measured. It is then soaked flat in a shallow dish for 2 h in water at 15–20 °C containing 0.5 g/l of an efficient wetting agent. It is removed and blotted dry with paper towels without unnecessary handling and allowed to dry flat at 20 °C on a smooth flat surface. It is then conditioned until equilibrium is reached and remeasured as described above.

The mean percentage change in each direction is calculated:

$$\text{Relaxation shrinkage} = \frac{\left(\begin{array}{c}\text{original measurement} - \\ \text{final measurement}\end{array}\right)}{\text{original measurement}} \times 100\%$$

6.2.4 Washable wool

When testing washable wool products for shrinkage it is usual to carry out tests that separate any felting shrinkage from relaxation shrinkage. It is important that the contribution of each type of shrinkage to the overall shrinkage is determined because both the cause and remedy for each type are quite different.

Relaxation and consolidation shrinkage

The US Standard [9] is for knitted fabrics containing at least 50% wool and which are designed to be shrink resistant. Three sets of marks are made in each direction each pair 10 in (254 mm) apart and each mark is situated at least 1 in (25 mm) from all edges.

The relaxation shrinkage is first determined by soaking the sample for 4 h at 38 °C, hydroextracting it, drying it flat at 60 °C, conditioning it and then measuring it.

The consolidation shrinkage is determined after relaxation shrinkage by agitating the sample in the Cubex apparatus for 5 min. The sample is placed in the cubex apparatus with 25 l of buffer solution at pH 7 containing 0.5% non-ionic detergent. The total load in the machine is 1 kg which is made up of the specimens plus makeweights, and the temperature of the solution is set to 40 °C. The sample is mechanically agitated for 5 min. It is then removed, rinsed three times and hydro-extracted followed by drying flat at 60 °C. Finally it is conditioned and remeasured.

The consolidation shrinkage value is calculated from the difference between the total shrinkage and the relaxation shrinkage.

IWS method

In this test method the relaxation shrinkage is determined from a wet treatment with mild agitation (standard 7A programme [10]). The felting shrinkage is determined subsequently on the same sample using a more severe agitation (standard 5A programme [10]) possibly using a number of repeat cycles.

Yarn and combed sliver can be tested by making them into a single jersey fabric.

Marking out

Knitted samples of size 300 mm × 400 mm are tested double with the free edges sewn together. The marks are placed not less than 25 mm from the edge.

Woven samples of size 500 mm × 500 mm have two of their edges sewn over after ironing as shown in Fig. 6.7 in order to test edge felting. Marks are made on the folded edge as well as in flat area of the sample. Shrinkproof treated fabrics are particularly vulnerable at folded over edges because of the more severe mechanical action in this region.

When testing socks three separate measurements should be made.

Measurement

Before measurement the samples are dried at a temperature not greater than 60 °C and then conditioned in the standard atmosphere for not less than 4 h.

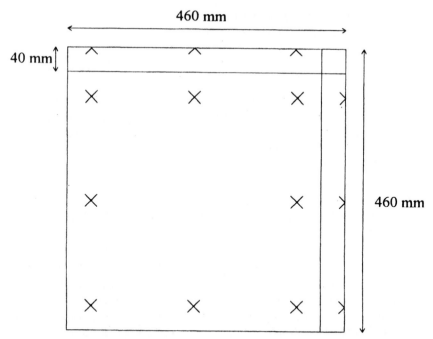

6.7   Marking out a woven sample for washable wool shrinkage test.

The measurements made are:

Original measurement	OM
After relaxation	RM
After felting	FM

Relaxation of the sample is accomplished by subjecting it to one cycle of the 7A programme in the Wascator washing machine (see below). The load is made up to a total of 1 kg with makeweights. The detergent concentration used is 0.3 g/l of the standard detergent SM49 and the washing is carried out at a temperature of 40 °C.

Felting of the sample is accomplished by washing it a number of times using the 5A programme in the Wascator, the actual number of washes depending on the product involved. The load, detergent and temperature used are the same as above.

Calculations

$$\text{Relaxation shrinkage} = \frac{(OM - RM)}{OM} \times 100\%$$

$$\text{Felting shrinkage} = \frac{(RM - FM)}{RM} \times 100\%$$

Cuff edge felting is unacceptable when the difference between the shrinkage of the fold and the flat area exceeds 1%.

Cubex

This test [11] is designed to measure the dimensional changes of wool-containing knitted fabrics during washing. It measures both relaxation and felting shrinkage as in the above test. Instead of a washing machine the test uses a cube-shaped drum of capacity 50 l which is suspended by opposite corners. The drum revolves at 60 revolutions per minute and the action is reversed every 5 min. The cube is filled through a small door in one face and there is no built-in heating mechanism. The knitted specimens are folded to give a double thickness 300 mm × 400 mm and the free edges are sewn together.

The test is carried out in two steps: first the relaxation shrinkage is measured and secondly any felting shrinkage is measured. The sample is first conditioned and measured flat. It is then placed in the cubex apparatus with 25 l of buffer solution at pH 7 containing 0.05% wetting agent. The load is 1 kg made up of specimens plus makeweights and the temperature of the solution is 40 °C. The sample is first given a static soak of 15 min which is followed by mechanical agitation for 5 min. It is then removed, rinsed three times and hydro-extracted followed by being dried flat at a temperature not above 55 °C. Finally it is conditioned and then remeasured.

In order to determine the felting shrinkage the same conditions are used but the agitation is continued for either 30 or 60 min depending on the severity of treatment required. The sample is dried, conditioned and remeasured as before.

The relaxation shrinkage, felting shrinkage and total shrinkage are calculated as in the previous test.

6.2.5 Washing programmes

Most tests for dimensional change due to washing use the procedures given in BS 4923 or ISO 6330 [10]. These standards give in detail the washing procedures for programmable washing machines. The reason that these details need to be specified is that a number of factors affect the intensity of the mechanical action of a rotary drum washing machine [12] such as the peripheral speed of the rotating drum, the height of the liquor in the drum, the liquor to goods ratio and the number and form of the lifters, in particular the height of them. Therefore a standard washing machine has to be used because the amount of agitation during washing has a bearing on the amount of shrinkage produced, particularly with wool. However, the programmes used in the machine are intended to be similar to the programmes found in domestic washing machines. The temperature and severity of the washing cycle used are also related to any care label that may be fixed to a garment made from the fabric being tested. In essence a fabric has to be able to undergo any laundering treatment recommended on the label without suffering from excessive dimensional change.

The standards specify the level of agitation during heating, washing and rinsing, the washing temperature, the liquor level during washing and rinsing, the washing time, whether there is a cool down after washing and the number of rinses and spin time. There are two sets of programmes in the standard, one for front loading machines designated type A and one for top loading agitator machines designated type B. Therefore a type 5A wash specifies a number 5 wash (40 °C, normal agitation) for a front loading machine. The use of a standard detergent without additives is also specified. The Wascator is an industrial washing machine which is commonly used for these tests and it can be programmed for all the main functions such as temperature, liquor level during washing and rinsing, washing time, time and number of rinses as required by the standard.

6.2.6   Dimensional stability to dry cleaning

The British Standard method [13] requires the use of a commercial dry cleaning machine. In the test the sample is prepared and marked out according to BS 4931 [6]. The total load used is 50 kg for each cubic metre of the machine cage made up of specimen plus makeweights. The solvent to be used is tetrachloroethylene containing 1 g/l of surfactant in a water emulsion, 6.5 litres of solvent being used for each kilogram of load. The machine is run for 15 min at 30 °C, the sample is rinsed in solvent and then dried by tumbling in warm air. The sample is then given an appropriate finishing treatment, which in most cases will be steam pressing, and it is then reconditioned and measured again.

6.2.7   Dimensional stability to dry heat

This test is intended to predict the behaviour of fabrics when heated in a hot press such as those used in various garment manufacturing processes including fusing and transfer printing. The ISO method [14] recommends that the samples are preconditioned at a low relative humidity before conditioning in the standard atmosphere. The samples are then marked out as shown in Fig. 6.8 and the dimensions AB, CD, EF and GH determined. They are then placed in a press heated to 150 °C under a pressure of 0.3 kPa for 20 s. The samples are conditioned and measured again so that the dimensional change in each direction can be calculated.

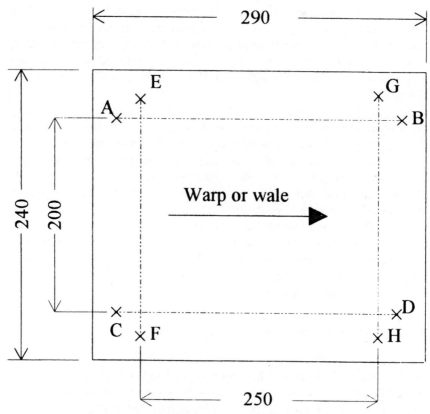

6.8   Marking out a sample for dimensional stability to dry heat.

General reading

Shaw T, 'The dimensional stability of woven wool fabrics', *Wool Sci Rev*, 1978 **55** 43.

References

1. Anon, 'Consistent objectives required', *Text Horizons*, 1989 **9** 47
2. Anon, 'Hygral expansion in wool fabrics', *Wool Sci Rev*, 1967 **31** 12.
3. Cookson P G, 'Relationships between hygral expansion, relaxation shrinkage, and extensibility in woven wool fabrics', *Text Res J*, 1992 **62** 44.
4. Abbott N J, Khoury F and Barish L, 'The mechanism of fabric shrinkage: the role of fibre swelling', *J Text Inst*, 1964 **55** T111–T127.
5. Anon, 'Geometry and dimensional properties of all wool weft knitted structures', *Wool Sci Rev*, 1971 **40** 14.
6. BS 4931 Preparation, marking and measuring of textile fabrics, garments and fabric assemblies in tests for assessing dimensional change.

7. BS 4323 Method for determination of dimensional change of fabrics induced by free steam.

8. ISO 7771 Determination of dimensional changes of fabrics induced by cold-water immersion.

9. ASTM D 1284 Test method for relaxation and consolidation dimensional changes of stabilised knit wool fabrics.

10. BS 4923 Individual domestic washing and drying for use in textile testing.

11. BS 1955 Method for determination of dimensional changes of wool-containing fabrics during washing.

12. Anon, 'The measurement of washing shrinkage in wool goods', *Wool Sci Rev*, 1970 **38** 35.

13. BS 4961 Determination of dimensional stability of textiles to dry cleaning in tetrachloroethylene.

14. ISO 9866 Effect of dry heat on fabrics under low pressure.

7.1 Introduction

A garment is considered to be serviceable when it is fit for its particular end use. After being used for a certain length of time the garment ceases to be serviceable when it can no longer fill its intended purpose in the way that it did when it was new. The particular factors that reduce the service life of a garment are heavily dependent on its end use. For instance overalls worn to protect clothing at work would be required to withstand a good deal of hard usage during their lifetime but their appearance would not be considered important. However, garments worn purely for their fashionable appearance are not required to be hard wearing but would be speedily discarded if their appearance changed noticeably. An exception to this generalisation is found in the case of denim where a worn appearance is deliberately strived for.

If asked, many people would equate the ability of a fabric to 'wear well' with its abrasion resistance, but 'wear', that is the reduction in serviceable life, is a complex phenomenon and can be brought about by any of the following factors:

1 Changes in fashion which mean that the garment is no longer worn whatever its physical state.
2 Shrinkage or other dimensional changes of such a magnitude that the garment will no longer fit.
3 Changes in the surface appearance of the fabric which include: the formation of shiny areas by rubbing, the formation of pills or surface fuzz, the pulling out of threads in the form of snags.
4 Fading of the colour of the garment through washing or exposure to light. The bleeding of the colour from one area to another.
5 Failure of the seams of the garment by breaking of the sewing thread or by seam slippage.

6 Wearing of the fabric into holes or wearing away of the surface finish or pile to leave the fabric threadbare. Wearing of the edges of cuffs, collars and other folded edges to give a frayed appearance.
7 Tearing of the fabric through being snagged by a sharp object.

These changes are brought about by the exposure of the garment to a number of physical and chemical agents during the course of its use. Some of these agents are as follows:

1 Abrasion of the fabric by rubbing against parts of the body or external surfaces.
2 The cutting action of grit particles which may be ingrained in dirty fabrics and which may cause internal abrasion as the fabric is flexed.
3 Tensile stresses and strains which occur as the garment is put on or taken off and when the person wearing it is active.
4 The laundering and cleaning processes which are necessary to retain the appearance of the garment.
5 Attack by biological agents such as bacteria, fungi and insects. This is a particular problem for natural materials.
6 Degradation of the fabric by contact with chemicals which can include normal household items such as bleach, detergents, anti-perspirants and perfumes.
7 Light, in particular ultra-violet light, can cause degradation of polymers leading to a reduction in strength as well as causing fading of colours.
8 Contact of the garment with sharp objects leading to the formation of tears.

The above causes of wear are often acting at the same time. For instance, chemical or bacterial attack may so weaken a fabric that it can then easily fail through abrasion or tearing. Laundering of a fabric taken together with the abrasion that it encounters during use may lead to much earlier formation of pills or failure through abrasion than would be predicted from any pilling or abrasion tests undertaken on the new material.

7.2   Snagging

A snag is a loop of fibre that is pulled from a fabric when it is in contact with a rough object. Snags detract from the appearance of the fabric but do not reduce any of its other properties. Fabrics made from bulked continuous filament yarns are particularly susceptible to the formation of snags although woven fabrics with long floats can also suffer from this problem.

7.2.1 Mace snagging test

The mace snagging test [1] is a comparative test for the snagging propensity of knitted fabrics of textured polyester yarn originally developed by ICI to test Crimplene yarns. In the test a metal ball fitted with spikes bounces randomly against a sleeve of the test fabric as it rotates. The spikes only catch loops of thread that are lying in a particular orientation so that it is important to test both directions of a fabric.

Four specimens, each one measuring 203 mm × 330 mm are tested; two with their long direction aligned with the length of the fabric and two with their long direction aligned with the fabric width. A seam is marked on the back of the fabric 16 mm from the shorter edge. The fabric samples are then folded face to face and sewn along the seam to form a tube. The tube is turned inside out so that the face of the fabric is on the outside. It is then slid over the cylinder of the machine and secured at each end with a rubber ring.

A mace is placed on each of the four fabric samples so that the chain holding it passes around the guide rod as shown in Fig. 7.1 and 7.2. The machine is then set to run for 600 revolutions (10 min).

When the test is complete the surface appearance of the specimen is compared with a set of photographic standards and given a rating from 5 (no snagging) to 1 (severe snagging).

7.3   Pilling

Pilling is a condition that arises in wear due to the formation of little 'pills' of entangled fibre clinging to the fabric surface giving it an unsightly appearance. Pills are formed by a rubbing action on loose fibres which are present on the fabric surface. Pilling was originally a fault found mainly in knitted woollen goods made from soft twisted yarns. The introduction of man-made fibres into clothing has aggravated its seriousness. The explanation for this is that these fibres are stronger than wool so that the pills remain attached to the fabric surface rather than breaking away as would be the case with wool. Figure 7.3 shows a pill on a cotton/polyester fabric.

The initial effect of abrasion on the surface of a fabric is the formation of fuzz as the result of two processes, the brushing up of free fibre ends not enclosed within the yarn structure and the conversion of fibre loops into free fibre ends by the pulling out of one of the two ends of the loop.

Gintis and Mead [2] consider that the fuzz formation must reach a critical height, which is dependent on fibre characteristics, before pill formation can occur.

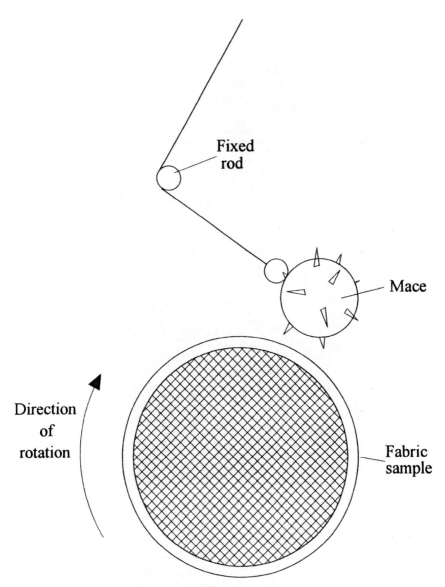

7.1   The mace snagging test.

The greater the breaking strength and the lower the bending stiffness of the fibres, the more likely they are to be pulled out of the fabric structure producing long protruding fibres. Fibre with low breaking strength and high bending stiffness will tend to break before being pulled fully out of the structure leading to shorter protruding fibres.

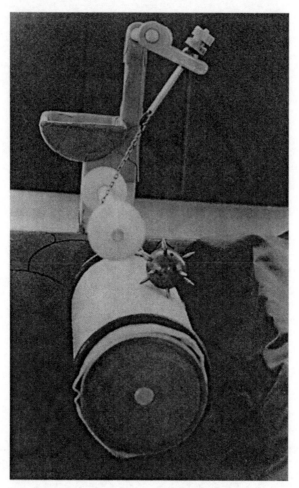

7.2   One station of a mace snagging tester.

The next stage is the entanglement of the loose fibres and the formation of them into a roughly spherical mass of fibres which is held to the surface by anchor fibres. As the pill undergoes further rubbing, the anchor fibres can be pulled further out of the structure or fatigued and eventually fractured depending on the fibre properties and how tightly they are held by the structure. In the case of low-strength fibres the pills will easily be detached from the fabric but with fabrics made from high-strength fibres the pills will tend to remain in place. This factor is responsible for the increase in the propensity for fabrics to pill with the introduction of synthetic fibres.

Low twist factors and loose fabric structures such as knitwear have a rapid fibre pull-out rate and long staple length resulting in the development

7.3   A pill ×50.

of numerous large pills. The life of these pills depends on the balance between the rate of fibre fatigue and the rate of roll-up. Pill density can either increase steadily, reach a plateau or pass through a maximum and decrease with time depending on the relative rates of pill formation and pill detachment. The pill density is also governed by the number of loose fibre ends on the surface and this may set an upper limit to the number of pills that will potentially develop. This has important implications for the length of a pilling test because if the test is carried on too long the pill density may have passed its maximum. Fibres with reduced flex life will increase the rate of pill wear-off.

Because the fibres that make up the pills come from the yarns in the fabric any changes which hold the fibres more firmly in the yarns will reduce the amount of pilling. The use of higher twist in the yarn, reduced yarn hairiness, longer fibres, increased inter-fibre friction, increased linear density of the fibre, brushing and cropping of the fabric surface to remove loose fibre ends, a high number of threads per unit length and special chemical treatments to reduce fibre migration will reduce the tendency to pill. The presence of softeners or fibre lubricants on a fabric will increase pilling. Fabrics made from blended fibres often have a greater tendency to pill as it has been found [3] that the finer fibres in a blend preferentially migrate towards the yarn exterior due to the difference in properties.

The amount of pilling that appears on a specific fabric in actual wear will vary with the individual wearer and the general conditions of use. Consequently garments made from the same fabric will show a wide range of pilling after wear which is much greater than that shown by replicate fabric specimens subjected to controlled laboratory tests.

Finishes and fabric surface changes may exert a large effect on pilling. Therefore, with some fabrics, it may be desirable to test before as well as after laundering or dry cleaning or both.

7.3.1 Pilling tests

After rubbing of a fabric it is possible to assess the amount of pilling quantitatively either by counting the number of pills or by removing and weighing them. However, pills observed in worn garments vary in size and appearance as well as in number. The appearance depends on the presence of lint in the pills or the degree of colour contrast with the ground fabric. These factors are not evaluated if the pilling is rated solely on the number or size of pills. Furthermore the development of pills is often accompanied by other surface changes such as the development of fuzz which affect the overall acceptability of a fabric. It is therefore desirable that fabrics tested in the laboratory are assessed subjectively with regard to their acceptability and not rated solely on the number of pills developed. Counting the pills and/or weighing them as a measure of pilling is very time consuming and there is also the difficulty of deciding which surface disturbances constitute pills. The more usual way of evaluation is to assess the pilling subjectively by comparing it with either standard samples or with photographs of them or by the use of a written scale of severity. Most scales are divided into five grades and run from grade 5, no pilling, to grade 1, very severe pilling.

ICI pilling box

For this test [4] four specimens each 125 mm × 125 mm are cut from the fabric. A seam allowance of 12 mm is marked on the back of each square. In two of the samples the seam is marked parallel to the warp direction and in the other two parallel to the weft direction. The samples are then folded face to face and a seam is sewn on the marked line. This gives two specimens with the seam parallel to the warp and two with the seam parallel to the weft. Each specimen is turned inside out and 6 mm cut off each end of it thus removing any sewing distortion. The fabric tubes made are then mounted on rubber tubes so that the length of tube showing at each end is the same. Each of the loose ends is taped with poly (vinyl chloride) (PVC) tape so that 6 mm of the rubber tube is left exposed as shown in Fig. 7.4. All four specimens are then placed in one pilling box. The samples are then

Table 7.1 Pilling grades

Rating	Description	Points to be taken into consideration
5	No change	No visual change
4	Slight change	Slight surface fuzzing
3	Moderate change	The specimen may exhibit one or both of the following: (a) moderate fuzzing (b) isolated fully formed pills
2	Significant change	Distinct fuzzing and/or pilling
1	Severe change	Dense fuzzing and/or pilling which covers the specimen.

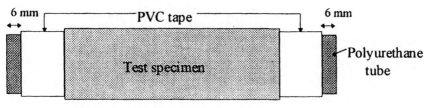

7.4   The preparation of a pilling sample.

tumbled together in a cork-lined box as shown in Fig. 7.5. The usual number of revolutions used in the test is 18,000 which takes 5 h. Some specifications require the test to be run for a different number of revolutions.

Assessment

The specimens are removed from the tubes and viewed using oblique lighting in order to throw the pills into relief. The samples are then given a rating of between 1 and 5 with the help of the descriptions in Table 7.1.

Random tumble pilling test

In this test [5] fabric specimens are subjected to a random rubbing motion produced by tumbling specimens in a cylindrical test chamber lined with a mildly abrasive material. In order to form pills that resemble those produced in actual wear in appearance and structure, small amounts of grey cotton lint are added to each test chamber with the specimens.

Three samples each 105 mm square are cut at an angle of 45° to the length of the fabric. The edges of the fabric samples are sealed by a suitable rubber adhesive to stop them fraying. All three samples are then placed in one test chamber which has been fitted with a fresh cork liner and 25 mg of the cotton lint is added. The machine is run for 30 min periods during which

7.5   A pilling box.

time the samples are tumbled by an impeller in the centre of the chamber. After each 30 min cycle the fabric is assessed and the chamber cleaned out and loaded with a fresh supply of lint. The number and timing of the cycles depend on the type of fabric being tested and would be laid down in the relevant specification. Figure 7.6 shows the chambers of a random tumble pilling tester.

In order to assess the amount of pilling on the fabrics they are placed in a suitable viewing cabinet which illuminates the pilled surface with light at a low angle so throwing the pills into relief. The fabric samples are assessed by comparing them with a set of photographic standards (ASTM or other), the rating being a subjective one using the following scale:

5 – no pilling
4 – slight pilling
3 – moderate pilling
2 – severe pilling
1 – very severe pilling

Pilling test Swiss standard

This test [6] uses the standard Martindale abrasion tester which is fitted with special large specimen holders and also has the driving pegs fitted at

7.6 The random tumble pilling tester.

a smaller radius in order to give a reduced specimen movement. The specimen holders are shown in Fig. 7.7 alongside the standard abrasion holder for comparison. The specimen under test is rubbed against a sample of the same fabric at a low pressure and then assessed for pilling in the normal way.

Two pressures are used depending on the type of fabric being tested:

- 6.5 cN/cm² woven and upholstery fabrics   extra weight in holder
- 2.5 cN/cm² knitted fabrics   holder only

Method

In the test three pairs of samples each 140 mm in diameter are cut from the fabric. One sample of each pair is mounted on the lower holder of the Martindale in place of the standard abradant. The other sample is mounted in the special pilling holder with a felt pad underneath. The sample is held in place by a large O ring. The sample holder is then mounted on the

7.7   A Swiss pilling holder compared with a standard holder.

machine in the normal way using a spindle but with no weight on the top of the spindle. Woven and upholstery fabrics require an extra circular weight placed in the top of the sample holder.

Three pairs of samples are tested as follows:

- One pair for 125 rubs.
- One pair for 500 rubs.
- One pair for 4 × 500 rubs, the specimens being brushed every 500 rubs to remove loose material. This pair forms the main assessment.

Assessment

Each pair of specimens is assessed and the grade is noted against the number of rubs although the final pair constitutes the main assessment.

Pilling is graded on a 5-point scale. If the degree of pilling is different on the upper and lower specimens then the upper specimen is assessed:

Grade 5 No or very weak formation of pills
Grade 4 Weak formation of pills
Grade 3 Moderate formation of pills
Grade 2 Obvious formation of pills
Grade 1 Severe formation of pills

Other Martindale pilling tests

A number of pilling tests have been designed around the standard Martindale abrasion tester. In most of these the fabric under test is mounted both in the holder and on the baseplate so that it is rubbed against itself. The fabric from the holder is the one that is usually assessed. Most test methods use the bare spindle without added weights but they differ in the number of rubs given to the sample. The results can then be assessed against a set of photographic standards. The advantage of these methods is that they are much quicker than the pill box.

7.4   Abrasion resistance

7.4.1 Factors affecting abrasion resistance

The evidence concerning the various factors that influence the abrasion resistance of fabrics is contradictory. This is because the experiments have been carried out under widely different conditions in particular using different modes of abrasion. Therefore the results are not comparable and often opposing results have been reported. The factors that have been found to affect abrasion [7, 8] include the following.

Fibre type

It is thought that the ability of a fibre to withstand repeated distortion is the key to its abrasion resistance. Therefore high elongation, elastic recovery and work of rupture are considered to be more important factors for a good degree of abrasion resistance in a fibre than is a high strength.

Nylon is generally considered to have the best abrasion resistance. Polyester and polypropylene are also considered to have good abrasion resistance. Blending either nylon or polyester with wool and cotton is found to increase their abrasion resistance at the expense of other properties. Acrylic and modacrylic have a lower resistance than these fibres while wool, cotton and high wet modulus viscose have a moderate abrasion resistance. Viscose and acetates are found to have the lowest degree of resistance to abrasion. However, synthetic fibres are produced in many different versions so that the abrasion resistance of a particular variant may not conform to the general ranking of fibres.

Fibre properties

One of the results of abrasion is the gradual removal of fibres from the yarns. Therefore factors that affect the cohesion of yarns will influence their abrasion resistance. Longer fibres incorporated into a fabric confer better abrasion resistance than short fibres because they are harder to remove from the yarn. For the same reason filament yarns are more abrasion resistant than staple yarns made from the same fibre. Increasing fibre diameter up to a limit improves abrasion resistance. Above the limit the increasing strains encountered in bending counteract any further advantage and also a decrease in the number of fibres in the cross-section lowers the fibre cohesion.

Yarn twist

There has been found to be an optimum amount of twist in a yarn to give the best abrasion resistance. At low-twist factors fibres can easily be removed from the yarn so that it is gradually reduced in diameter. At high-twist levels the fibres are held more tightly but the yarn is stiffer so it is unable to flatten or distort under pressure when being abraded. It is this ability to distort that enables the yarn to resist abrasion.

Abrasion resistance is also reported to increase with increasing linear density at constant fabric mass per unit area.

Fabric structure

The crimp of the yarns in the fabric affects whether the warp or the weft is abraded the most. Fabrics with the crimp evenly distributed between warp and weft give the best wear because the damage is spread evenly between them. If one set of yarns is predominantly on the surface then this set will wear most; this effect can be used to protect the load-bearing yarns preferentially. One set of yarns can also be protected by using floats in the other set such as in a sateen or twill weave. The relative mobility of the floats helps to absorb the stress.

There is an optimum value for fabric sett for best abrasion resistance. The more threads per centimetre there are in a fabric, the less force each individual thread has to take. However, as the threads become jammed together they are then unable to deflect under load and thus absorb the distortion.

7.4.2 Abrasion tests

Factors affecting abrasion tests

Very many different abrasion tests have been introduced [7, 8]. Poor correlation has been found both between the different abrasion testers and between abrasion tests and wear tests [8, 9]. The methods that have survived to become standards are not necessarily the 'best' ones. Among the factors which can affect the results of an abrasion test are the following.

Type of abrasion

This may be plane, flex or edge abrasion or a combination of more than one of these factors.

Type of abradant

A number of different abradants have been used in abrasion tests including standard fabrics, steel plates and abrasive paper or stones (aluminium oxide or silicon carbide). The severity as well as the type of action is different in each case. For the test to correspond with actual wear in use it is desirable that the abrasive should be similar to that encountered in service. An important concern is that the action of the abradant should be constant throughout the test. It is likely that the abradant itself will wear during the test thus changing its abrasive properties. Equally it can become coated in material from the abraded sample, such as finishes which can then act as lubricants so reducing its effectiveness.

Pressure

The pressure between the abradant and the sample affects the severity and rate at which abrasion occurs. It has been shown that using different pressures can seriously alter the ranking of a set of fabrics when using a particular abradant [8]. Accelerated destruction of test samples through increased pressure or other factors may lead to false conclusions on fabric behaviour. For instance accelerated tests do not allow for any relaxation of fibres and fabrics a factor which can be expected during normal use.

Speed

Increasing the speed of rubbing above that found in everyday use also brings the dangers of accelerated testing as described above. A rise in temperature of the sample can occur with high rubbing speeds; this can affect the physical properties of thermoplastic fibres.

Tension

It is important that the tension of the mounted specimen is reproducible as this determines the degree of mobility of the sample under the applied abradant. This includes the compressibility of any backing foam or inflated diaphragm.

Direction of abrasion

In many fabrics the abrasion resistance in the warp direction differs from that of the weft direction. Ideally the rubbing motion used by an abrasion machine should be such as to eliminate directional effects.

Method of assessment

Two approaches have been used to assess the effects of abrasion:

1   Abrade the sample until a predetermined end-point such as a hole, and record the time or number of cycles to this.
2   Abrade for a set time or number of cycles and assess some aspect of the abraded fabric such as change in appearance, loss of mass, loss of strength change in thickness or other relevant property.

The first approach corresponds to most people's idea of the end point of abrasion but the length of the test is indeterminate and requires the sample to be regularly examined for failure in the absence of a suitable automatic mechanism. This need for examination is time consuming as the test may last for a long time. The second approach promises a more precise measurement but even when the sample has rubbed into a hole the change in properties such as mass loss can be slight.

However none of the above assessment methods produces results that show a linear or direct comparison with one another [8]. Neither is there a linear relationship between successive measurements using any of these methods and progressive amounts of abrasion.

Martindale

This apparatus [10] is designed to give a controlled amount of abrasion between fabric surfaces at comparatively low pressures in continuously changing directions.

The results of this test should not be used indiscriminately, particularly not for comparing fabrics of widely different fibre composition or construction.

In the test circular specimens are abraded under known pressure on an apparatus, shown in Fig. 7.8, which gives a motion that is the resultant of two simple harmonic motions at right angles to one another. The fabric under test is abraded against a standard fabric. Resistance to abrasion is estimated by visual appearance or by loss in mass of the specimen.

Method

Four specimens each 38mm in diameter are cut using the appropriate cutter. They are then mounted in the specimen holders with a circle of standard foam behind the fabric being tested. The components of the standard holder are shown in Fig. 7.9. It is important that the mounting of the sample is carried out with the specimens placed flat against the mounting block.

7.8   The Martindale abrasion tester.

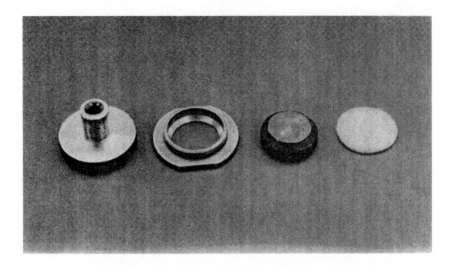

7.9   A standard holder for the Martindale abrasion test.

The test specimen holders are mounted on the machine with the fabric under test next to the abradant. A spindle is inserted through the top plate and the correct weight (usually of a size to give a pressure of 12 kPa but a lower pressure of 9 kPa may be used if specified) is placed on top of this. Figure 7.10 shows the sample mounted in a holder. The standard abradant should be replaced at the start of each test and after 50,000 cycles if the test is continued beyond this number. While the abradant is being replaced it is

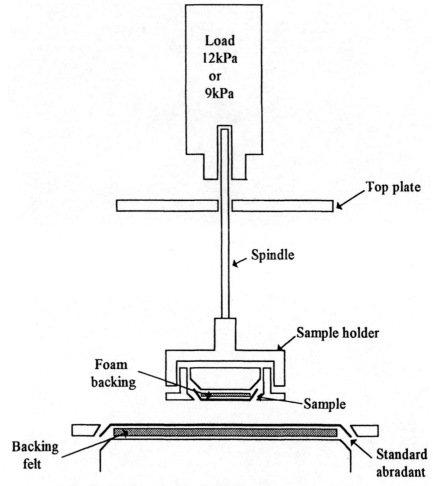

7.10   One station of a Martindale abrasion tester.

held flat by a weight as the retaining ring is tightened. Behind the abradant is a standard backing felt which is replaced at longer intervals.

Assessment

The specimen is examined at suitable intervals without removing it from its holder to see whether two threads are broken. See Table 7.2 for the time lapse between examinations. If the likely failure point is known the first inspection can be made at 60% of that value. The abrading is continued until two threads are broken. All four specimens should be judged individually.

Table 7.2 Inspection intervals for Martindale abrasion test

Estimated number of cycles	Intervals for inspection
Up to 5,000	Every 1,000
Between 5,000 and 20,000	Every 2,000
Between 20,000 and 40,000	Every 5,000
Above 40,000	Every 10,000

The individual values of cycles to breakdown of all four specimens are reported and also the average of these.

Average rate of loss in mass

This is an alternative method of assessing abrasion resistance which requires eight specimens for the test. Two of these are abraded to the end-point as described above and then the other pairs are abraded to the inter-mediate stages of 25%, 50% and 75% of the end point. The samples are weighed to the nearest 1 mg before and after abrasion so that a graph can be plotted of weight loss against the number of rubs. From the slope of this graph, if it is a straight line, the average loss in mass measured in mg/1000 rubs can be determined.

Abrasion resistance for hosiery

This test makes use of a modified specimen holder for the Martindale abra-sion tester, which stretches the knitted material thus effectively accelerat-ing the test. The holder, shown in Fig. 7.11, takes a standard size 38 mm diameter sample which is held to size by a pinned ring. A flattened rub-ber ball is pushed through the sample as the holder is tightened thus stretch-ing it. The holder is then mounted on the Martindale with a 12 kPa weight and the test carried out as normal. The sample is inspected at suitable inter-vals until a hole appears or the material develops an unacceptable level of thinning.

Accelerotor

The Accelerotor abrasion tester [11] has an action that is quite different from most other abrasion testers. In the test an unfettered fabric specimen is driven by rotor inside a circular chamber lined with an abrasive cloth.

7.11   A sock abrasion test holder.

The apparatus, shown in Fig. 7.12, is fitted with a variable speed drive and a tachometer to indicate the rotation speed. The sample suffers abrasion by rubbing against itself as well as the liner. Evaluation is made either on the basis of the weight loss of the sample or on the loss in grab strength of the specimen broken at an abraded edge. In each case three specimens are tested.

For evaluation by weight loss, square specimens are cut with pinking shears, the size of specimen being determined by the cloth weight. The cut edges are coated with adhesive to prevent fraying and allowed to dry. The specimens are conditioned and then weighed to ±0.001 g. Each specimen is then placed in the Accelerotor and run for the desired time at the selected speed. It is then taken out from the machine, any loose debris removed, conditioned and weighed again. The percentage weight loss for each specimen is then calculated.

For evaluation by loss in strength, specimens measuring 100 mm × 300 mm are used in order to provide two grab test samples. Each specimen is numbered at both ends and then cut in half. One half is used for determining the original grab strength and the other half for determining the grab strength after abrading. The half to be abraded has its edges adhesive coated as above. It is then folded 50 mm from the short edge making it into a 100 mm² square. This flap is stitched down to the main body so that the folded edge will be abraded during the test. The sample is then run in the Accelerotor under the chosen conditions and the stitching removed.

7.12 The Accelerotor abrasion tester.

The breaking strength is then determined by the grab method (described in Chapter 5), making sure that the worn edge is in the portion being tested. The breaking strength of the matching original is also determined and the percentage loss in breaking strength of each pair of specimens is calculated.

Taber abraser

In this instrument [12] the fabric is subjected to the wear action of two abrasive wheels which press onto a rotating sample. The wheels are arranged at diametrically opposite sides of the sample so that they are rotated in the opposite direction by the rotation of the sample. The abrading wheels travel on the material about a horizontal axis which is displaced tangentially from the axis of the test material, so resulting in a sliding action of the abrasive on the sample. This gives rise to an X pattern of wear caused by the tracks of the two abrasive wheels being displaced relative to each other. Debris from the abrasive action is removed during the test by a vacuum nozzle.

The wheels normally used for testing textiles are the rubber base resilient type composed of abrasive grains embedded in rubber. These are made in different abrasive grain sizes. The loads used can be 125, 250, 500 or 1000 gf (1.23, 2.45, 4.9 or 9.81 N).

Evaluation can be by: (1) the number of cycles to a visual end-point, that is a predetermined point at which the material has undergone a marked change in appearance such as removal of the pile or when it has broken down physically; (2) residual breaking load; the breaking load of the abraded sample is measured using a gauge length of 25 mm (1 in), making sure that the abraded part of the sample is between the jaws; or (3) percentage loss in breaking load, obtained by calculating the breaking load after abrasion as a percentage of the breaking load of the original fabric.

7.5   Wearer trials

The main purpose of laboratory tests is to obtain prior knowledge of the performance of textile products in service. The assumption is made that when such tests are carried out, there is some relationship between the results of the laboratory tests and the performance of the items in use. In order to design laboratory testing procedures that correlate with end use performance the conditions of actual use must be carefully analysed so that they can be simulated as closely as possible in a controlled setting. Since actual wear is such a complex phenomenon, however, laboratory tests are usually designed to evaluate only one or a limited number of variables at a time.

In a wearer trial the product (garment, furnishing, carpet etc.) is used in the 'normal' manner and a report is made at intervals on its behaviour. When comparing these trials with laboratory testing there are certain important differences.

In general user trials are not widely used in industry. Wearer trials are more usually carried out by large organisations, for example BTTG, IWS, Courtaulds and ICI, very often using their own staff as garment users. The trials are often used to compare a new material or process against one that is known to be satisfactory in service. The cost of user trials may be very high and as a result they are most used for fairly low-priced but common articles, for example, socks, tights, tea-towels, shirts, children's trousers, blazers and sheets, rather than for carpets or furniture.

7.5.1 Advantages of wearer trials

1   In a wearer trial the material receives treatment similar to that experienced in normal wear. For example clothing breaks down due to a combination of loading, flexing, pilling and rubbing together with the effect of light, perspiration and bacteria. These causes can interact to produce a more rapid breakdown than would be the case with any of the indi-

vidual causes. It may not be possible to imitate the normal wear pattern in a laboratory.

2   A wearer trial tests all the components which make up a garment such as buttons, sewing thread, seams, lining and cuffs. Laboratory tests on the separate components may not show faults due to making up.

7.5.2 Disadvantages of wearer trials

1   Wearer trials are difficult to control and organise as it is necessary to rely on the user to treat the article normally and to report accurately and at the required time on its performance. It is quite possible that in a large trial some garments may be untraced at the end of the work because of people moving, losing interest or the article may become lost or destroyed.

2   Wearer trials are expensive because of the cost of producing garments from a fabric rather than testing the fabric itself. Parallel trials may also be needed using control garments if, for instance, an improved product is to be compared with a standard product. There are also the personnel costs to be considered in collecting, assessing and distributing articles.

3   It is impossible to achieve 'normal wear'. The intensity of wear depends on a large number of factors: the type of employment of the wearer, for instance manual workers may put more strain on their clothes than office workers, and the cleanliness of the working environment also plays a part; the size and weight of the person, in that a large person may be expected to put a higher stress on certain parts of a garment, and closeness of fit of the garment is a related factor; the individual habits of the wearers, for instance they may ride a bicycle to work so causing extra wear on trousers; the time of year – items such as pullovers will be worn more often in the cold months. The weather also influences the wearing of other garments at the same time as the test garment so having an effect on the pilling performance of the test garment for instance.

4   One of the main problems in conducting wearer trials is that of finding suitable groups of people who live similar lives, come together on a regular basis and who will co-operate. It is not possible to select 50 people at random from the phone book as they would never be seen again after handing out the garments. Suitable groups of people include: police officers, nurses, post-office staff, boarding school pupils, prisoners and students.

5   In a trial the garments are usually examined and assessed at regular intervals. These assessments cannot be destructive as the garments have to be worn again, so they have to be subjective. Ideally an individual trial would finish at some definite change in property such as the appear-

ance of a hole but the criteria for judging that the end of a garment's useful life has been reached are not usually as definite as this and involve a judgement as to what is unacceptable. Therefore there is a serious problem with the accuracy and reproducibility of these assessments and their relationship with laboratory tests.

6   The most serious problem with wearer trials is that they take a very long time to complete as their time span must be similar to that of the life expectancy of the article being tested and are therefore no use if rapid results are required.

7.5.3 Advantages of laboratory tests

1   They are rapid. Most tests can be completed within a day.
2   They are designed to give objective results. A numerical result or rating allows one fabric to be ranked as being better or worse than another fabric even when the differences between them are small.
3   The tests are under the direct control of the tester. This allows the conditions of test to be exactly specified and factors other than those under test to be kept constant.
4   They can be reproduced. An identical test carried out on the same fabric should ideally give the same result in any laboratory and with any operator.

7.5.4 Disadvantages of laboratory tests

1   Laboratory tests can only imitate wear conditions
2   For a complete evaluation of a fabric it is necessary to use a large range of expensive equipment.
3   Laboratory tests are rapid because many of them aim to accelerate the natural causes of wear. Speeding up a test may give false results, for example the continuous action of abrasion tests may cause heating of the material which is not present in normal use.

7.5.5 Design of trials

In planning a trial a balance has to be struck between what should be done and what can be done. It is convenient to issue garments for a certain number of days then collect, inspect and wash them all under identical conditions. This eliminates any possibility of differences in performance which are due to different washing conditions. Alternatively, users may be left to wash the garment in their normal way as well as wearing it, making out a report at intervals. This of course introduces further variables but may be considered closer to 'normal' use. It is important in such tests that the person wearing the garment should not know the details of composition, etc.

The US Standard for wearer trials [13] has the following recommendations: that control garments are used which have a known wear performance history. It is not possible to ensure that wearer trials undertaken at different times have the same severity as the people who undertake the trial and their circumstances can change. It is not possible to repeat a wearer trial as each one is different.

1   Decide on the garment that is to be tested.
2   Define the object of the test and the information that is to be obtained. For instance: the performance properties to be evaluated, the areas of garments to be examined, how the performance will be evaluated and what scale is to be used for this. Decide in advance what ratings for these properties will constitute satisfactory or unsatisfactory performance.
3   Establish the percentage of specimens that must fail in order to constitute overall unsatisfactory performance. The test is terminated when this point has been reached.
4   Establish the number of wash/wear cycles that will constitute satisfactory performance.
5   Define the wear – laundering cycle (or other method of refurbishing) by the number of hours worn or by the number of wearings before laundering and the method of laundering.
6   Decide on the number of participants.
7   Permanently label each garment with a code to identify the wearer and garment, keep new garment for comparison purposes.
8   Issue garments with instructions.
9   Evaluate after each wash/wear cycle and record ratings.

References

1. BS 3838 Specification for blazer cloths Appendix A.
2. Gintis D and Mead E J, 'The mechanism of pilling', *Text Res J*, 1959 **29** 578–585.
3. Anon. 'Methods and finishes for reducing pilling Part 1', *Wool Sci Rev*, 1972 **42** 32.
4. BS 5811 Method of test for determination of the resistance to pilling of woven fabrics (pill box method).
5. ASTM D3512 Pilling resistance and other related surface changes of textile fabrics: random tumble pilling tester method.
6. SN 198 525 Testing of textiles; testing of pilling resistance.
7. Galbraith R L, 'Abrasion of textile surfaces' in *Surface Characteristics of Fibres and Textiles*, part 1, Schick M J ed. Dekker Inc., New York, 1975.
8. Bird S L, *A Review of the Prediction of Textile Wear Performance with Specific Reference to Abrasion*, SAWTRI Special Publication, Port Elizabeth, 1984.

9. Committee of Directors of Textile Research Associations, 'Final report on inter-laboratory abrasion tests', *J Text Inst*, 1964 **55** P1.
10. BS 5690 Method of test for determination of the abrasion resistance of textiles.
11. AATCC 93 Abrasion resistance of fabrics: accelerotor method.
12. ASTM D3884 Abrasion resistance of textile fabrics (rotary platform, double head method).
13. ASTM D3181 Standard practice for conducting wear testing on textile garments.

8
Comfort

8.1 Introduction

It is possible to distinguish two aspects of wear comfort of clothing:

1 Thermophysiological wear comfort which concerns the heat and moisture transport properties of clothing and the way that clothing helps to maintain the heat balance of the body during various levels of activity.
2 Skin sensational wear comfort which concerns the mechanical contact of the fabric with the skin, its softness and pliability in movement and its lack of prickle, irritation and cling when damp.

8.2 Thermal comfort

8.2.1 Heat balance

The human body tries to maintain a constant core temperature of about 37 °C. The actual value varies slightly from person to person but the temperature of any one person is maintained within narrow limits. In most climates body temperature is above that of the external environment so that there has to be an internal source of heat in order to maintain the temperature difference. The required heat comes from the body's metabolism, that is the necessary burning of calories to provide power to the muscles and other internal functions. However, the body must be kept in thermal balance: the metabolic heat generated together with the heat received from external sources must be matched by the loss from the body of an equivalent amount of heat. If the heat gain and the heat loss are not in balance then the body temperature will either rise or fall, leading to a serious threat to life.

The efficiency of the human organism is such that of the energy taken in as food only 15–30% is converted into useful work with the remaining 70–85% of the energy being wasted as heat. Any level of physical activity

Table 8.1 Energy costs of activities [1]

Activity	Energy cost (watts)
Sleeping	70
Resting	90
Walking 1.6 km/h (1 mph)	140–175
Walking 4.8 km/h (3 mph)	280–350
Cycling 16 km/h (10 mph)	420–490
Hard physical work	445–545
Running 8 km/h (5 mph)	700–770
Sprinting	1400–1500

above that needed to maintain body temperature will result in an excess of heat energy which must be dissipated, otherwise the body temperature will rise. A lower level of physical activity will lead to a fall in body temperature if the available heat is not conserved by increased insulation.

The approximate energy costs which are associated with human activity are shown in Table 8.1 and range from a minimum value of about 70 W when sleeping to an absolute maximum of about 1500 W which is only possible in short bursts. A rate of about 500 W (corresponding to hard physical work) can be kept up for a number of hours.

If a person is comfortable (that is, in heat balance) at rest then a burst of hard exercise will mean that there is a large amount of excess heat and also perspiration to be dissipated. On the other hand if the person is in heat balance during strenuous exercise then he or she will feel cold when resting owing to the large reduction of heat generation.

8.2.2 Heat loss

There are four mechanisms that allow the body to lose heat to the environment in order to maintain its thermal balance. The way the heat loss is divided between the mechanisms depends on the external environment.

1 **Conduction.** In this process heat loss is accomplished through direct contact with another substance. The rate of exchange is determined by the temperature difference between the two substances and by their thermal conductivities. For example the body loses heat in this manner when submerged in cold water.

2 **Convection.** This is a process in which heat is transferred by a moving fluid (liquid or gas). For example, air in contact with the body is heated by conduction and is then carried away from the body by convection.

3 **Radiation.** This is the process of heat transfer by way of electromagnetic waves. The waves can pass through air without imparting much heat to

it; however, when they strike an object their energy is largely transformed into heat. Radiation can largely be ignored as a mechanism of losing heat as it is very dependent on the temperature of an object (varying as T^4) so that it is more important as a means of heat gain from very hot bodies such as the sun, radiant heaters or fires. Heat radiation and absorption by an object are both influenced by its colour. Black is both the best absorber and radiator of heat. White and polished metals are poor absorbers and radiators as most of the energy is reflected. Clothing acts to reduce radiation loss by reducing the temperature differences between the body and its immediate surroundings as the clothing effectively becomes the immediate surroundings.

4 **Evaporation.** Changing liquid water into vapour requires large amounts of heat energy. One calorie will raise the temperature of one gram of water one degree Celsius; however, it takes 2424 J (580 calories) to evaporate one gram of water at body temperature. If the water is evaporated from the skin surface then the energy required is removed from the skin, thus cooling it. When environmental temperatures approach skin temperature (35 °C seated to 30 °C heavy physical work), heat loss through convection and radiation gradually come to an end so that at environmental temperatures above skin temperature the only means for the body to lose heat is through evaporation of sweat. Sweating itself is not effective as it is the conversion of the liquid to vapour that removes the heat. This mechanism works well in a hot dry environment but evaporation of sweat becomes a problem in hot humid climates.

The requirements for heat balance vary with the climate; in hot climates the problem is one of heat dissipation whereas in cold climates it is one of heat conservation.

Clothing has a large part to play in the maintenance of heat balance as it modifies the heat loss from the skin surface and at the same time has the secondary effect of altering the moisture loss from the skin. However, no one clothing system is suitable for all occasions: a clothing system that is suitable for one climate is usually completely unsuitable for another.

The main fabric properties that are of importance for maintaining thermal comfort are:

- insulation,
- windproofing,
- moisture vapour permeability,
- waterproofing.

These properties are closely interrelated in that changes in one property can cause changes to the other three properties.

Insulation

An air temperature of 28–29 °C would be required for a person to be able to sit in comfort without wearing any clothes. At air temperatures lower than this, therefore, the body will lose heat without the added insulation given by clothing.

If losses by convection can be prevented, the air itself offers a very high resistance to heat conduction having a value of thermal resistance which is only slightly less than that of a vacuum. Convection losses arise because the body loses heat to the air in contact with it. This heated air is then immediately replaced with cooler air either through natural convection or through air currents. The air currents can be caused by either body movement or by external air flow such as in windy conditions. Convection losses can therefore be reduced by keeping the air surrounding the body at rest. Air tends to 'cling' to solid surfaces so that material with a large exposed surface area, such as a mass of fine fibres, acts as a good restrictor of air movement. In clothing the majority of the bulk is composed of air, for example a worsted suiting fabric is made up of 25% fibre and 75% air whereas knitted underwear and quilted fabrics filled with fibre battings or down and feathers may contain 10% or less actual fibre, with the rest consisting of air.

The heat flow through a fabric is due to a combination of conduction and radiation the convection within a fabric being negligible [2]. The conduction loss is determined by the thickness of the fabric and its thermal conductivity. The thermal conductivity is itself a combination of the conductivity of air k_A and that of the fibre k_F:

$$\text{Fabric conductivity } k = (1 - f)k_A + fk_F$$

where f is the fraction by volume of the fabric taken up by fibre. The conductivity of air is 0.025 W/m K and that of fibres is 0.1 W/m K, therefore in fabrics of fibre contents below 10% the conductivity is effectively that of air.

The heat flow due to radiation is more complex as it is governed by the temperature difference between the heat emitter and the heat absorber. The infra-red radiation only travels a few millimetres into a fabric as it is either scattered or absorbed by the fibres. These fibres in turn emit radiation which travels a further short distance to the next fibres and so on until it reaches the far surface. Therefore the radiative heat transfer between the body and the external environment is indirect and depends on the absorption and emission properties of the fibres.

In order to predict the heat flow due to radiation through a fabric it is necessary to know the temperature profile. The simplest assumption is a linear change in temperature with distance through the fabric which is true of conduction heat flow. At the edges of a fabric the situation is more complex than this, but in the centre of a thick specimen the conductivity due to radiation can be simplified to [2]:

$$\text{Radiative conductivity} = \frac{8\sigma T^3 R}{f\varepsilon}$$

where σ = Stefan–Boltzmann constant ($5.67 \times 10^{-8}\,\text{W/m}^2\,\text{K}^4$),
  ε = thermal emissivity,
  R = radius of fibres,
  T = mean temperature between heat source and sink (K),
  f = fractional fibre volume.

This means that the heat loss from radiation is higher at low fibre volumes (less than 5%) but is reduced by the use of fine fibres and higher fibre volumes [2].

The insulation value of clothing when it is worn is not just dependent on the insulation value of each individual garment but on the whole outfit as the air gaps between the layers of clothing can add considerably to the total insulation value. This assumes that the gaps are not so large that air movement can take place within them, leading to heat loss by convection. Because of this limitation the closeness of fit of a garment has a great influence on its insulation value as well as the fabric from which it is constructed.

The insulation value of a fabric is in fact mainly dependent on its thickness and it can be estimated from the relationship:

$$\text{clo} = 1.6 \times \text{thickness in cm}$$

where clo is a measure of thermal resistance (see below).

Unfortunately any increase in the thickness of clothing in order to increase its insulation value also increases its surface area and bulk and thus decreases its rate of water vapour transmission.

The insulation value of clothing is generally measured by its thermal resistance which is the reciprocal of thermal conductivity. The advantage of using thermal resistance rather than conductivity is that the values of the different layers of clothing can then be added together to give the overall resistance value of the outfit. The thickness of the material is not taken into consideration in the units of thermal resistance used. The SI unit of thermal resistance is degrees Kelvin square metre per watt ($1\,\text{unit} = 1\,\text{K}\,\text{m}^2/\text{W}$). Two different units are in use in the clothing field:

Clo unit

This is an American unit which was adopted for the simplicity of presentation to a layperson. Clothing having a thermal resistance of one clo unit approximates the normal indoor clothing worn by sedentary workers, i.e. suit, shirt and underwear.

The complete definition is the insulation required to maintain a resting man producing heat at 50 Kcal/m^2/h indefinitely comfortable in an atmosphere at 21 °C less than 50% RH and air movement of 10 cm/s. Its value is equivalent to 0.155 °C m^2/W.

Tog unit

This is a British unit which was proposed by workers at the Shirley Institute and it is related to the international units for thermal resistance used for building materials but converted to larger units to make them easier to use:

$$1 \text{ tog} = 0.1\,°\text{C}\,\text{m}^2/\text{W}$$
$$10 \text{ togs} = 1\,\text{K}\,\text{m}^2/\text{W} = 1\,°\text{C}\,\text{m}^2/\text{W}$$
$$1 \text{ clo} = 1.55 \text{ tog}$$

Windproofing

Air tends to cling to surfaces and thus form an insulating layer. The value of this can be as high as 0.85 clo at a windspeed of 0.15 m/s [3] even on an uncovered surface. This is the reason that the comfort temperature for the unclothed body is lower than the skin temperature. However, the air layer is easily disturbed by air movements; Fig. 8.1 shows the fall in its insulation value with increasing windspeed which is governed by the following equation [3]:

$$\text{Insulation of surface layer} = \frac{1}{0.61 + 0.19V^{1/2}} \text{ clo}$$

where V = air velocity cm/s.

Wind has two other major effects on clothing: firstly it causes compression of the underlying fabrics and therefore reduces their insulation value by reducing their thickness. Secondly it disturbs the trapped air in the clothing system by both the movement of the fabric and by penetration through the fabric, thus increasing the convection heat losses. At low windspeeds (less than 8 km/h) these effects are negligible [1] but at windspeeds of 32–48 km/h they can become significant depending on the air permeability of the outer layer of the clothing. For example an impermeable cover fabric worn over a pile insulating layer showed an 18% loss in insulation at a windspeed of 48 km/h due to compression. With the use of permeable outer fabrics the convection losses were:

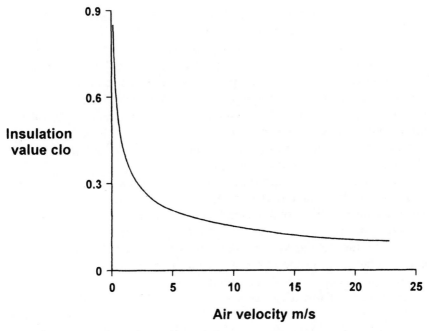

8.1   The insulation values of air.

10%   4.5 oz (125 g) downproof sateen
43%   5.5 oz (155 g) cotton gabardine
73%   3 oz (85 g) poplin shirting

A fabric can easily be made windproof by coating it with an impermeable coating or by using a tight weave. However, coated fabrics have low
rate of moisture vapour transmission.

Moisture vapour permeability

Perspiration is an important mechanism which the body uses to lose heat
when its temperature starts to rise. Heat is taken from the body in order to
supply the latent heat needed to evaporate the moisture from the skin.

There are two forms of perspiration:

1   Insensible – in this form the perspiration is transported as a vapour and
it passes through the air gaps between yarns in a fabric.
2   Liquid – this form occurs at higher sweating rates and it wets the clothing which is in contact with the skin.

The two forms of perspiration raise separate problems: one is the ability of
water vapour to pass through the fabric, particularly the outer layer; and

the other is the ability of the fabric in contact with the skin of absorbing or otherwise dealing with the liquid sweat.

The ability of a fabric to allow perspiration in its vapour form to pass through it is measured by its moisture vapour permeability in grams of water vapour per square metre per 24 hours. A fabric of low moisture vapour permeability is unable to pass sufficient perspiration and this leads to sweat accumulation in the clothing and hence discomfort. The fabrics most likely to have a low permeability are the ones that have been coated to make them waterproof. The coatings used to keep out liquid water will also block the transport of water vapour.

The overall moisture vapour permeability of clothing is rather like the thermal insulation value in that it depends on the whole clothing system. The resistance of each clothing item together with air gaps adds to the total resistance of the system. If the production of perspiration is greater than the amount the clothing system will allow to escape, moisture will accumulate at some point in the clothing system. If the outer layer is the most impermeable, moisture will accumulate in the inner layers. When excess moisture accumulates it causes a reduction in thermal insulation of the clothing and eventually condensation and wetting. The level of perspiration production is very dependent on the level of activity: clothing that may be comfortable at low levels of activity may be unable to pass sufficient moisture vapour during vigorous activity. However, when activity ceases, freezing can occur because the clothing is now damp and body heat production has been reduced, leading to after-exercise chill and, if the temperature is low enough, frostbite. This factor is particularly important in Arctic conditions where excessive sweating must be avoided at all costs. The overall combination of wind, wetting and disturbance of entrapped air can reduce the nominal insulation value of a clothing system by as much as 90%.

Waterproof breathable fabrics

The existence of the problem of the lack of water vapour permeability in waterproof fabrics has led to the development of waterproof breathable fabrics. One approach to producing such a fabric is to use membranes attached to the fabric or coatings on the fabric which are waterproof but which will allow moisture vapour to pass through. It is possible to achieve this because of the enormous difference in size between a water droplet and a water vapour molecule. A water droplet has a size of around $100\,\mu m$ whereas a water vapour molecule has a size of around $0.0004\,\mu m$. If, therefore, a membrane or fabric can be produced with pore sizes between these two limits it will then have the desired properties.

Another approach is to use a coating or membrane made from a hydrophilic film without any pores. The moisture is absorbed by the membrane and then transported across it by diffusion, evaporating when it reaches the outer surface. The driving force for the diffusion is provided by the body temperature which ensures that the water passes out even when it is raining.

A third approach is to use a tightly woven fabric. One of the original waterproof breathable fabrics, Ventile, was produced in this way by using tightly woven special cotton yarns which swelled when wet thus closing any gaps in the fabric. The production of fine microfibres has allowed the production of fabrics that are woven sufficiently tightly to keep liquid water out.

Lightweight polyurethane-coated nylon fabrics typically have moisture permeability values of the order of 200 g/m²/24 h, whereas the requirement when walking at 5 km/h is that the fabric should pass between 2800 and 7000 g/m²/24 h (average surface area of human body 1.7–1.8 m²). Cotton ventile fabrics have moisture permeability values of about 4900 g/m²/24 h whereas modern membrane materials have quoted values of between 4000 g/m²/24 h and 2500 g/m²/24 h [4]; the comparability of these figures depends on the way the values have been measured.

Overall there is a trade-off between the degree of waterproofing of a fabric and its moisture vapour transmission.

Waterproofing

Waterproofing is very important for the outer layer of a clothing system designed to be worn outdoors. This property is particularly important in cold weather activities for keeping the insulation of any clothing system dry. Waterlogging of fabrics fills up the air spaces with water and hence reduces their insulation value considerably as shown in Table 8.2. If the water penetrates to the skin it can also remove a large amount of heat by the same mechanism as that which makes perspiration effective.

The waterproofing of fabrics can readily be achieved by the use of synthetic polymer coatings; however, the use of simple coatings bring with it the penalty of excess build-up of moisture vapour above certain levels of activity. The design of clothing for comfort and protection in adverse weather conditions is therefore a matter of compromise between the competing requirements. No one fabric or clothing item can fulfil all the requirements, the clothing system as a whole has to be considered.

8.2.3 Air permeability

The air permeability of a fabric is a measure of how well it allows the passage of air through it. The ease or otherwise of passage of air is of importance for a number of fabric end uses such as industrial filters, tents,

Table 8.2 Effect of moisture on insulation values [5]

	Thermal insulation (togs)		
	Staple polyester	Continuous filament polyester	50/50 down/feather
Dry	4.7	6.8	6.6
15% moisture	2.4	3.6	2.8
50% moisture	1.8	2.4	1.5
100% moisture	1.7	2.3	1.3

sailcloths, parachutes, raincoat materials, shirtings, downproof fabrics and airbags.

Air permeability is defined as the volume of air in millilitres which is passed in one second through $100 \, s \, mm^2$ of the fabric at a pressure difference of 10 mm head of water.

In the British Standard test [6] the airflow through a given area of fabric is measured at a constant pressure drop across the fabric of 10 mm head of water. The specimen is clamped over the air inlet of the apparatus with the use of rubber gaskets and air is sucked through it by means of a pump as shown in Fig. 8.2. The air valve is adjusted to give a pressure drop across the fabric of 10 mm head of water and the air flow is then measured using a flowmeter.

Five specimens are used each with a test area of $508 \, mm^2$ (25.4 mm diameter) and the mean air flow in ml per second is calculated from the five results. From this the air permeability can be calculated in ml per $100 \, mm^2$ per second.

The reciprocal of air permeability, air resistance, can be defined as the time in seconds for 1 ml of air to pass through $100 \, s \, mm^2$ of fabric under a pressure head of 10 mm of water. The advantage of using air resistance instead of air permeability to characterise a fabric is that in an assembly of a number of fabrics, the total air resistance is then the sum of the individual air resistances.

To obtain accurate results in the test, edge leakage around the specimen has to be prevented by using a guard ring or similar device (for example, efficient clamping). The pressure drop across the guard ring is measured by a separate pressure gauge. Air that is drawn through the guard ring does not pass through the flowmeter. The pressure drops across the guard ring and test area are equalised in order that no air can pass either way through the edge of the specimen. A guard ring of three times the size of the test area is considered sufficient.

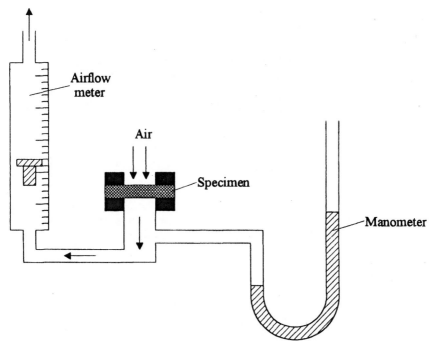

8.2   The air permeability test.

8.2.4 Measurement of thermal conductivity

The transmission of heat through a fabric occurs both by conduction through the fibre and the entrapped air and by radiation. Practical methods of test for thermal conductivity measure the total heat transmitted by both mechanisms. The insulation value of a fabric is measured by its thermal resistance which is the reciprocal of thermal conductivity (transmittance) and it is defined as the ratio of the temperature difference between the two faces of the fabric to the rate of flow of heat per unit area normal to the faces. As can be seen from this definition it is necessary to know the rate of heat flow through a fabric in order to be able to measure its thermal resistance. In practice the measurement of the rate of heat flow in a particular direction is difficult as a heater, even when supplied with a known amount of power, dissipates its heat in all directions. Two different methods are in use to overcome this problem: one is to compare thermal resistance of the sample with that of a known standard and the other is to eliminate any loss in heat other than that which passes through the fabric being tested. It is important that any measurements of thermal resistance are made at temperatures close to those that are likely to be encountered in use as the thermal conductivity of materials varies with the temperature. This is due

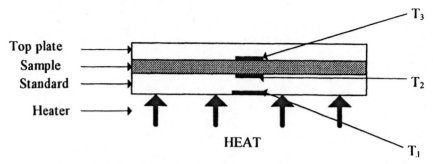

8.3 Togmeter: two plate method.

to the variation in thermal conductivity of the air with temperature and also the dependence of the heat loss through radiation on temperature.

Togmeter

The togmeter [7] avoids the problem of measuring heat flow by placing a material of known thermal resistance in series with the material under test so that the heat flow is the same through both materials. The thermal resistance of the test fabric can then be calculated by comparing the temperature drop across it with the temperature drop across the standard material.

Apparatus

The togmeter consists of a thermostatically controlled heating plate which is covered with a layer of insulating board of known thermal resistance. The temperature is measured at both faces of this standard. The heater is adjusted so that the temperature of the upper face of the standard is at skin temperature (31–35 °C). A small airflow is maintained over the apparatus.

There are two methods of test that can be used with the togmeter:

1 **Two plate method.** In this method the specimen under test is placed between the heated lower plate and an insulated top plate as shown in Fig. 8.3. The top plate has a low mass so that it does not compress the fabric. The temperature is measured at the heater (T_1), between the standard and the test fabric (T_2) and between the fabric and the top plate (T_3).

2 **Single plate method.** In this method the specimen under test is placed on the heated lower plate as above but it is left uncovered as shown in Fig. 8.4, the top plate being used to measure the air temperature (T_3).

The air above the test specimen has a considerable thermal resistance itself so that the method is in fact measuring the sum of the specimen

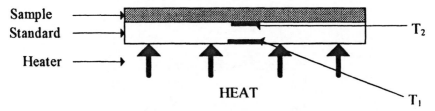

Sample

Standard

Heater

HEAT

T$_2$

T$_1$

8.4   Togmeter: single plate method.

thermal resistance and the air thermal resistance. A separate experiment is therefore performed without the specimen (i.e. a bare-plate test) to measure the resistance of the air R_{air}.

To determine the air resistance

The heater and the fan are switched on and the apparatus is allowed to reach thermal equilibrium with no specimen present. The top plate is placed underneath the apparatus shielded from radiation by a foil-covered plate, in order to measure the air temperature. The temperature should remain steady at each thermocouple for 30 mins. It may take some time for an equilibrium to be reached. Thermal resistance of air:

$$R_{air} = R_{stand} \times \frac{T_2 - T_3}{T_1 - T_2}$$

where R_{stand} is the thermal resistance of the standard.

To determine the sample resistance

The above experiment is repeated with the test sample placed on the bottom plate and the apparatus again allowed to reach thermal equilibrium. Thermal resistance of sample:

$$R_{sample} = R_{stand} \times \frac{T_2 - T_3}{T_1 - T_2} - R_{air}$$

Guarded hotplate method

The guarded hotplate [8] is used to measure thermal transmittance which is the reciprocal of the thermal resistance. The apparatus consists of a heated test plate surrounded by a guard ring and with a bottom plate underneath as shown in Fig. 8.5. All three plates consist of heating elements sand-

Top view

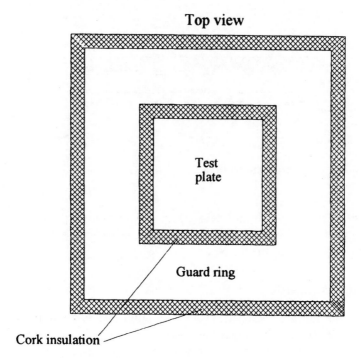

Test
plate

Guard ring

Cork insulation

Fabric sample

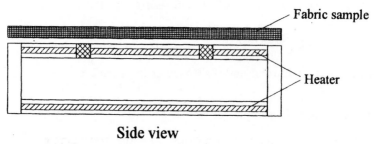

Heater

Side view

8.5 The guarded hotplate.

wiched between aluminium sheets. All the plates are maintained at the same constant temperature in the range of human skin temperature (33–36 °C). The guard ring and bottom plate, which are maintained at the same temperature as the test plate, ensure that no heat is lost apart from that which passes upwards through the fabric under test. The whole apparatus is covered by a hood to give still air conditions around the specimen. The whole of the surroundings of the apparatus is maintained at fixed conditions between 4.5 and 21.1 °C and 20 and 80% RH, the exact conditions being specified as part of the test.

With the test fabric in place the apparatus is allowed to reach equilibrium before any readings are taken. This may take some time with thick specimens. The amount of heat passing through the sample in watts per square metre is measured from the power consumption of the test plate heater. The temperature of the test plate and the air 500 mm above the test plate are measured.

The measured thermal transmittance consists of the thermal transmittance of the fabric plus the thermal transmittance of the air layer above the fabric which is not negligible. Therefore the test is repeated without any fabric sample present to give the bare plate transmittance. The transmittance of the air layer above the plate is assumed to be the same as that of the air layer above the sample.

Combined transmittance of specimen and air U_1:

$$U_1 = \frac{P}{A \times (T_p - T_a)} \, \text{W/m}^2 \, \text{K}$$

where: P = power loss from test plate (W),
$\quad\quad A$ = area of test plate (m^2),
$\quad\quad T_p$ = test plate temperature (°C),
$\quad\quad T_a$ = air temperature (°C).

The bare plate transmittance U_{bp} is similarly calculated and then the intrinsic transmittance of the fabric alone, U_2, is calculated from the following equation:

$$\frac{1}{U_2} = \frac{1}{U_1} - \frac{1}{U_{bp}}$$

8.2.5   Measurement of water vapour permeability

The water vapour permeability of fabrics is an important property for those used in clothing systems intended to be worn during vigorous activity. The human body cools itself by sweat production and evaporation during periods of high activity. The clothing must be able to remove this moisture in order to maintain comfort and reduce the degradation of thermal insulation caused by moisture build-up. This is an important factor in cold environments.

The main materials of interest are those fabrics that incorporate a polymer layer that makes the fabric waterproof but which still allows some water vapour to pass through. There are two main types of these materials: those that contain pores through which the moisture vapour can pass and those containing a continuous layer of hydrophilic polymer. The mechanism of water vapour transmission through the second type is quite different from that of the first type. In particular the rate of diffusion through the hydrophilic polymer layer is dependent on the concentration of water vapour in the layer. The higher the concentration, the higher the rate of transfer. In the materials where transmission is via pores the rate is independent of water vapour concentration. This has a bearing on the results obtained from the different methods of testing water vapour permeability from the two types of material which can rank them differently depending on the test method used.

Cup method

In the British Standard version of this method [9] the specimen under test is sealed over the open mouth of a dish containing water and placed in the standard testing atmosphere. After a period of time to establish equilibrium, successive weighings of the dish are made and the rate of water vapour transfer through the specimen is calculated.

The water vapour permeability index is calculated by expressing the water vapour permeability (WVP) of the fabric as a percentage of the WVP of a reference fabric which is tested alongside the test specimen.

Each dish is filled with sufficient distilled water to give a 10mm air gap between the water surface and the fabric. A wire sample support is placed on each dish to keep the fabric level. Contact adhesive is applied to the rim of the dish and the specimen, which is 96mm in diameter, is carefully placed on top with its outside surface uppermost. The cover ring is then placed over the dish and the gap between cover ring and dish sealed with PVC tape as shown in Fig. 8.6.

A dish which is covered with the reference fabric is also set up in the same way. All the dishes are then placed in the standard atmosphere and allowed to stand for at least 1 h to establish equilibrium.

Each dish is then weighed to the nearest 0.001 g and the time noted. After a suitable time for example overnight the dishes are reweighed and the time noted again.

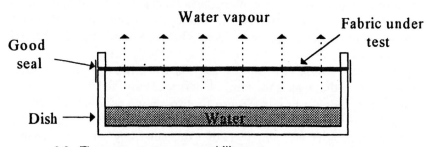

8.6   The water vapour permeability test.

Calculate:

$$\text{WVP} = \frac{24M}{At}\, \text{g}/\text{m}^2/\text{day}$$

where: M = loss in mass (g),
   t = time between weighings (h),
   A = internal area of dish (m^2).

$$A = \frac{\pi d^2 \times 10^{-6}}{4}$$

where d = internal diameter of dish (mm).

$$\text{Water vapour permeability index} = \frac{(\text{WVP})_f \times 100}{(\text{WVP})_r}$$

where WVP$_f$ is the water vapour permeability of the test fabric and WVP$_r$ is the water vapour permeability of the reference fabric.

The ASTM method E 96–80 [10] procedure B is similar to the above method although the air gap above the water surface is 19 mm (0.75 in) and an air velocity of 2.8 m/s (550 ft/min) is used over the surface of the fabric.

The airgaps above the specimen are important with these tests as the air itself has a high resistance to water vapour permeability [11]. Figure 8.7 shows that the total resistance to water vapour permeability of the experimental set-up depends on three factors.

The experiment is sometimes carried out with the cup inverted so that the water is in contact with the inner surface of the fabric [11]. This form of the test tends to give more favourable results for hydrophilic films.

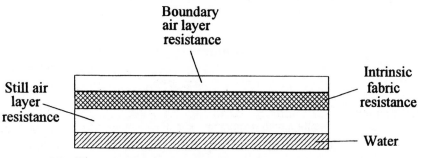

8.7   The various resistances to water vapour permeability.

Sweating guarded hotplate method

An alternative method to the cup method is to use a plate that is heated to skin temperature and supplied with water in order to simulate sweating. This is much closer to actual conditions of use than the cup method but it requires a more sophisticated experimental procedure. A number of methods have been described that differ in the way of supplying the water to the fabric and in the way of measuring the water vapour passing through it.

The sweating guarded hotplate [11] is similar to the guarded hotplate which is used to measure thermal resistance. In the normal test the power required to maintain the plate at a given temperature is related to the dry thermal resistance of the material. If the plate is saturated with water the power required is then related to the rate at which water evaporates from the surface of the plate and diffuses through the material in addition to the dry thermal resistance.

In order to measure the water vapour permeability of a material, therefore, it is first necessary to measure the dry thermal transmittance U_1 as described in section 8.2.2 on measuring thermal conductivity. The measurement is then repeated with the plate supplied with moisture. This is achieved by using a saturated porous plate covered with a Cellophane film, as shown in Fig. 8.8, which allows moisture to pass through but not in sufficient quantity to wick into the fabric. A moisture vapour permeability index i_m is calculated from the following formula:

$$i_m = \frac{\left[\dfrac{PR_{tot}}{A}\right] - (T_p - T_a)}{S(p_s - \phi p_a)}$$

where: R_{tot} = $1/U_1$ = resistance of the fabric plus boundary air layer $(m^2 K/W)$,

A = surface area (m^2),

T_p = temperature of the saturated plate surface,

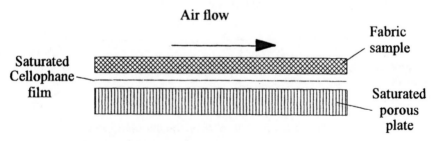

Air flow

Saturated
Cellophane
film

Fabric
sample

Saturated
porous
plate

8.8   The sweating hotplate.

T_a = temperature of the ambient air,

P = power required to maintain a constant saturated plate surface temperature (W),

S = Lewis relation between evaporative mass transfer coefficient and convective heat transfer coefficient (1.65×10^{-2}K/Pa),

p_s = saturated water vapour pressure at the plate surface (Pa),

p_a = saturated water vapour pressure of the ambient air (Pa),

ϕ = relative humidity of the ambient air.

The i_m value is a relative measure which should vary between 0 for completely impermeable materials and 1 for completely permeable materials.

The moisture permeability index i_m can be combined with the dry thermal resistance R_{tot} to give i_m/R_{tot} which is a measure of both evaporative heat flow and other forms of heat flow. The higher the value for i_m/R_{tot}, the better the material is at dissipating heat by all mechanisms.

The moisture vapour transmission rate (MVTR) for the sweating guarded hotplate is:

$$\mathrm{MVTR}_{plate} = 1.04 \times 10^3 \left(\frac{i_m}{R_{tot}}\right) g/m^2/24\,h$$

8.3   Moisture transport

In order to keep the wearer dry and hence comfortable, clothing that is worn during vigorous activity, such as sports clothing, has to be able to deal with the perspiration produced by such activity. There are two main properties of clothing, that affect the handling of moisture. Firstly there is the ease with which clothing allows the perspiration to be evaporated from the skin surface during the activity. Secondly after the activity has ceased, there is a need for the moisture that is contained in the clothing layer next to the skin to dry out quickly. This ensures that the wearer does not lose heat unnecessarily through having a wet skin. Some workers [12] also consider that the extent to which the wet fabric clings to the skin is also important to the comfort of a garment.

Moisture is transmitted through fabrics in two ways:

1   By diffusion of water vapour through the fabric. This appears to be independent of fibre type but is governed by the fabric structure. The measurement of air flow through the fabric provides a good guide to its ability to pass water vapour in large quantities.

2   By the wicking of liquid water away from the skin using the mechanism of capillary transport. The ability of a fabric to do this is dependent on the surface properties of the constituent fibres and their total surface

area. The size and number of the capillary paths through the fabric are also very important but these are governed by factors such as the fibre size, the yarn structure and the fabric structure. The capillary network of the fabric is dependent on the direction under consideration so that the wicking properties through the thickness of the fabric may be different from those in the plane of the fabric. Also the rate of wicking may be different along the warp (wale) direction than along the weft (course) direction.

8.3.1 Wetting

For wicking to take place the fibre has first to be wet by the liquid. In fact it is the balance of forces involved in wetting the fibre surface that drives the wicking process. When a fibre is wetted by a liquid the existing fibre–air interface is displaced by a new fibre–liquid interface. The forces involved in the equilibrium that exists when a liquid is in contact with a solid and a vapour at the same time are given by the following equation:

$$\gamma_{SV} - \gamma_{SL} = \gamma_{LV} \cos \theta$$

where γ represents the interfacial tensions that exist between the various combinations of solid; liquid and vapour; the subscripts S, L and V standing for solid, liquid and vapour,

θ = equilibrium contact angle,
γ_{LV} = the surface tension of the liquid.

The contact angle is defined as the angle between the solid surface and the tangent to the water surface as it approaches the solid; the angle is shown as θ in Fig. 8.9. The angle is determined by the three interfacial tensions: if γ_{SV} is larger than γ_{SL} then $\cos \theta$ is positive and the contact angle must be between 0° and 90°. If γ_{SV} is smaller than γ_{SL} then the contact angle must be between 90° and 180°. A high contact angle for water with the surface means that water will run off it, a low contact angle means that

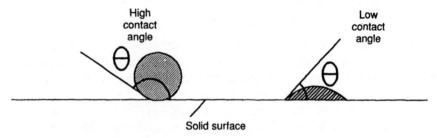

8.9 Contact angle.

water will wet the material. Water repellent materials exhibit a high contact angle. A contact angle of less than 90° also means that water will wick into the material by capillary action. A contact angle of 90° or more means that water will not rise by capillary action. The measured (apparent) contact angle shows hysteresis in that the contact angle for a liquid that is advancing is usually higher than that for a liquid that is receding. The advancing contact angle is usually used in wicking calculations.

8.3.2 Wicking

In the absence of external forces the transport of liquids into fibrous assemblies is driven by capillary forces that arise from the wetting of the fibre surfaces described above. If the liquid does not wet the fibres it will not wick into the fibrous assembly. In the case of contact angles above 90°, liquid in a capillary is depressed below the surface instead of rising above it. In order for the wicking process to take place spontaneously, the balance of energy has to be such that energy is gained as the liquid advances into the material, therefore γ_{SV} must be greater than γ_{SL}:

$$\text{Work of penetration, } W_p = \gamma_{SV} - \gamma_{SL} = \gamma_{LV} \cos\theta$$

The wicking rate is dependent on the capillary dimensions of the fibrous assembly and the viscosity of the liquid. For a simple capillary of radius r the rate of progress of the liquid front shown diagrammatically in Fig. 8.10 is given by:

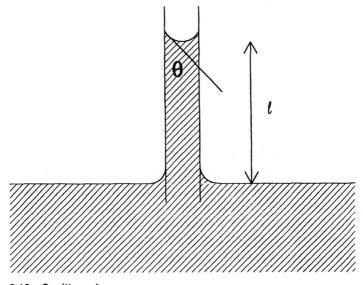

8.10 Capillary rise.

$$\frac{dl}{dt} = \frac{r\gamma_{LV}\cos\theta_A}{4\eta l}$$

where θ_A = advancing contact angle,

η = viscosity of liquid,

l = length of liquid front.

The wetting of fibres is purely dependent on their surface properties, in particular in the case of wetting with water, whether the surface is hydrophobic or hydrophilic. Therefore the wetting and wicking properties of fibres can be modified by surface finishes and experimental studies can also be affected by the remains of processing oils and finishes. Wetting is also affected by the presence of surfactants in the liquid which alter its interfacial tensions.

When wicking takes place in a material whose fibres can absorb liquid the fibres may swell as the liquid is taken up, so reducing the capillary spaces between fibres, potentially altering the rate of wicking.

8.3.3 Longitudinal wicking

The distance l travelled along a capillary by a liquid in time t is given by:

$$l = \left(\frac{rt\gamma_{LV}\cos\theta_A}{2\eta}\right)^{0.5}$$

If the material is vertical the height to which the liquid wicks is limited by gravitational forces and ceases when the capillary forces are balanced by the weight of liquid:

$$\text{Equilibrium height } l = \frac{2\gamma_{LV}\cos\theta_A}{rg\rho}$$

where ρ = liquid density,

g = gravitational acceleration.

8.3.4 Wicking test

In this test [13] a strip of fabric is suspended vertically with its lower edge in a reservoir of distilled water as shown in Fig. 8.11. The rate of rise of the leading edge of the water is then monitored. To detect the position of the water line a dye can be added to the water or, in the case of dark coloured fabrics, the conductivity of the water may be used to complete an electrical circuit. The measured height of rise in a given time is taken as a direct indication of the wickability of the test fabric.

The simple form of the test does not take into account the mass of the water that is taken up. This will depend on the height the water has risen

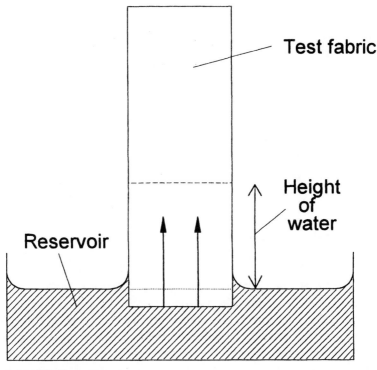

8.11 Wicking test.

to, the thickness of the fabric and the water-holding power of the fabric structure. One way of allowing for this is to weigh the fabric at the end of the test and hence obtain the mass of the water taken up by the fabric. The mass can then be expressed as a percentage of the mass of the length of dry fabric which is equivalent to the measured height of water rise.

8.3.5 Transverse wicking

Transverse wicking is the transmission of water through the thickness of a fabric, that is, perpendicular to the plane of the fabric. It is perhaps of more importance than longitudinal wicking because the mechanism of removal of liquid perspiration from the skin involves its movement through the fabric thickness. Transverse wicking is more difficult to measure than longitudinal wicking as the distances involved are very small and hence the time taken to traverse the thickness of the fabric is short.

One test is the plate test which consists of a horizontal sintered glass plate kept moist by a water supply whose height can be adjusted so as to keep the water level precisely at the upper surface of the plate. A fabric placed

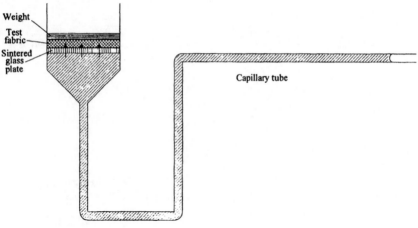

8.12   The plate test.

on top of this glass plate as shown in Fig. 8.12 can then draw water from the glass plate at a rate which depends on its wicking power. It is important that the water level is adjusted to touch the underside of the fabric but not to flood it. The rate of uptake of water is measured by timing the movement of the meniscus along the long horizontal capillary tube. The equipment is arranged so that the head of water supplying the glass plate does not change during the experiment. Given the diameter of the capillary tube, the mass of water taken up by the fabric in a given time can be calculated. A problem encountered with this method is that a load has to be placed on top of the fabric to ensure contact with the sintered glass plate. Unfortunately when a fabric is compressed its structural elements are moved closer to one another which can potentially change its absorption characteristics. A contact pressure of 0.098 kPa has been used by Harnett and Mehta [13].

8.4   Sensorial comfort

Sensorial comfort is concerned with how a fabric or garment feels when it is worn next to the skin. It has been found that when subjects wore various fabrics next to the skin they could not detect differences in fabric structure, drape or fabric finish but could detect differences in fabric hairiness [14]. Some of the separate factors contributing to sensorial comfort which have been identified [14] are:

1   **Tickle** caused by fabric hairiness.
2   **Prickle** caused by coarse and therefore stiff fibres protruding from fabric surface. Matsudaira *et al.* [15] found that the stiffness of protruding fibres

is the dominant factor in causing prickle sensations. This is affected by both fibre diameter and to a lesser extent fibre length. For a fibre of a given diameter the end of a long fibre is more easily deflected a fixed amount than the end of a short fibre so it appears less prickly. For a fibre of a given length a larger diameter is much stiffer depending on the fourth power of the diameter and hence is more likely to prick.

3 **Wet cling** which is associated with sweating and is caused by damp and sticky sweat residues on skin. A factor influencing cling is the actual area of fabric in contact with skin which in turn is influenced by fabric structure.

4 **Warmth to touch**: when a garment is first picked up or put on it is usually at a lower temperature than the skin and thus there will be a loss of heat from the body to the garment until the temperatures of the surfaces in contact equalise. The faster this heat transfer occurs, the greater is the cold feel of the fabric. The differences in cold feel between fabrics is mainly determined by their surface structure rather than by the fibre type.

For example a cotton sheet feels cool whereas a flannelette sheet, which is produced by raising the surface of a cotton fabric, feels comparatively warm. The raised surface gives a lower contact area and hence a slower rate of change of temperature. Ironing a cotton sheet has the effect of increasing the cold feel by compacting the surface structure.

8.5   Water absorption

Some textile end uses such as towels, tea towels, cleaning cloths, nappies (diapers) and incontinence pads require the material to absorb water. There are two facets to the absorption of water: one is the total amount that can be absorbed regardless of time and the other is the speed of uptake of the water. These two properties are not necessarily related as fabrics of similar structures but with different rates of uptake may ultimately hold similar amounts of water if enough time is allowed for them to reach equilibrium. Alternatively soaking the fabrics in water so that they take up their maximum load may mask any differences in rate of uptake.

8.5.1 Static immersion

The static immersion test [16] is a method for measuring the total amount of water that a fabric will absorb. Sufficient time is allowed in the test for the fabric to reach its equilibrium absorption.

In the test weighed samples of the fabric are immersed in water for a given length of time, taken out and the excess water removed by shaking.

They are then weighed again and the weight of water absorbed is calculated as a percentage of the dry weight of the fabric.

Four specimens each 80 mm × 80 mm are cut at 45° to the warp direction. The first step is to condition the samples and weigh them. They are then immersed in distilled water at a temperature of 20 ± 1 °C to a depth of 10 cm. A wire sinker is used to hold the specimens at the required depth. The samples are left in this position for 20 min. After the specimens are taken from the water the surface water is removed immediately from them by shaking them ten times in a mechanical shaker. They are then transferred directly to preweighed airtight containers and then reweighed:

$$\text{Absorption} = \frac{\text{mass of water absorbed}}{\text{original mass}} \times 100\%$$

The mean percentage absorption is calculated.

8.5.2 Wettability of textile fabrics

This is a test [17] for fabrics containing hydrophilic fibres. **Wettability** is defined as the time in seconds for a drop of water or 50% sugar solution to sink into the fabric. Fabrics that give times exceeding 200 s are considered unwettable.

In the test the specimen is clamped onto an embroidery frame 150 mm in diameter so that it is held taut and away from any surface. A burette with a standard tip size (specified in the standard) is clamped 6 mm above the horizontal surface of the sample as shown in Fig. 8.13. The fabric is illuminated at an angle of 45° and is viewed at 45° from the opposite direction so that any water on the surface reflects the light to the viewer. At the start of a test a drop of liquid is allowed to fall from the burette and the timer started. When the diffuse reflection from the liquid vanishes and the liquid

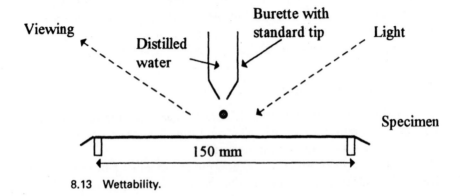

8.13 Wettability.

is no longer visible, the timing is stopped. If the time recorded is less than 2 s, the burette is changed to one containing 50% sugar solution.

Five areas on each specimen are tested, three samples in all, to give a total of 15 measurements.

8.5.3 Sinking time

This is a simple test for highly absorbent materials in which a 25 mm × 25 mm piece of fabric or a 50 mm length of yarn taken from the fabric is dropped onto the surface of distilled water and the length of time it takes to sink is measured. If the sample does not sink within 1 min it is considered as having floated.

8.6   Water repellency

A number of fabric end uses, particularly those where fabric is used out of doors, require the material to be more or less impermeable to rain. These include outerwear such as anoraks, cagoules and raincoats and also industrial fabrics such as tents and tarpaulins. Broadly two main categories of resistance to water penetration are recognised, based mainly on the treatment that has been used on the fabric:

1 *Waterproof.* A waterproof fabric is one that is coated or impregnated to form a continuous barrier to the passage of water using for example rubber, polyurethane, PVC or wax coatings. In such fabrics the gaps between the yarns are filled in by the coating which gives rise to two main drawbacks. Firstly the fabric will no longer allow water vapour to pass through it, making it uncomfortable to wear when sweating. Secondly the binding together of the yarns by the coating reduces the ability of the fabric to shear and thus to mould to the body contours.

2 *Showerproof.* Showerproof fabrics are ones that have been treated in a such a way as to delay the absorption and penetration of water. Showerproofing of fabrics is often achieved by coating them with a thin film of a hydrophobic compound such as a silicone. The film covers the surface of the individual fibres making them water repellent. When a fabric has been treated in this way a drop of water on the surface does not spread. The process leaves the gaps in the fabric weave untouched so keeping it quite permeable to air and water vapour. The process also leaves the handle of the fabric largely unaffected unlike fabrics with a waterproof coating. However, water can penetrate the fabric if it strikes it with sufficient force as in heavy rain; alternatively, the flexing of the fabric during wear can cause the gaps in the weave to open and close so allowing the water to penetrate.

A showerproof fabric can also be produced by a correct choice of yarn and fabric construction to give a very tight weave which physically keeps the water out; an example of this is the gabardine construction used in coats.

8.6.1 Spray rating

The spray rating test [18] is one used to measure the resistance of a fabric to surface wetting but not to penetration of water. It is therefore a test which is particularly used on showerproof finishes. It is often the case that waterproof coatings are applied to the inner surface of a material and a water repellent finish is then applied to the outer fabric surface to stop it absorbing water as it would otherwise become waterlogged. In such cases the test is used on the outer layer of fabrics which are otherwise considered waterproof.

In the test three specimens are tested, each one 180 mm square. Each specimen in turn is held taut over a 150 mm diameter embroidery hoop which is mounted at 45° to the horizontal. A funnel which is fitted with a standard nozzle containing 19 holes of a specified diameter is held 150 mm above the fabric surface as shown in Fig. 8.14. Into the funnel is poured 250 ml of distilled water at 20 °C to give a continuous shower onto the fabric. After the water spray has finished the hoop and specimen are removed and tapped twice smartly against a solid object on opposite points of the frame, the fabric being kept horizontal. This removes any large drops of water. The fabric is then assigned a spray rating either using the written grading shown in Table 8.3 or from photographic standards (American Association of Textile Chemists and Colorists scale).

8.6.2 Bundesmann water repellency test

The Bundesmann test aims to produce the effect of a rainstorm on a fabric in the laboratory. In the test shown in Fig. 8.15 the fabric is subjected to a shower of water from a head fitted with a large number of standard nozzles. During the shower the back of the fabric is rubbed by a special mechanism which is intended to simulate the flexing effect which takes place when the fabric is worn.

The method is not currently a British standard because considerable variation has been found between different machines, although when tests are carried out on the same machine the variability can be reduced to acceptable levels.

In the test four specimens are mounted over cups in which a spring loaded wiper rubs the back of the cloth while the whole cup assembly slowly rotates. They are subjected for 10 min to a heavy shower whose rate has been adjusted so as to deliver 65 ml of water per minute to each cup. The

Table 8.3 Spray ratings

Grade	Description
1	Complete wetting of the whole of the sprayed surface
2	Wetting of more than half the sprayed surface
3	Wetting of the sprayed surface only at small discrete areas
4	No wetting of but adherence of small drops to the sprayed surface
5	No wetting of and no adherence of small drops to the sprayed surface

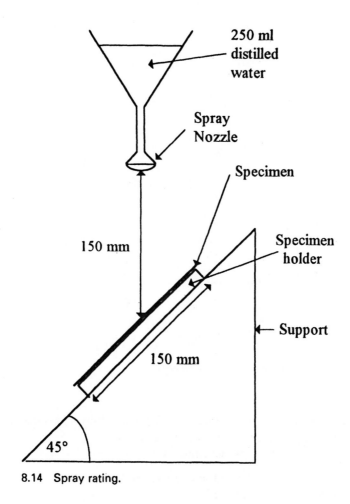

250 ml distilled water

Spray Nozzle

Specimen

Specimen holder

150 mm

Support

150 mm

45°

8.14   Spray rating.

water flow is maintained at 20 °C and between pH 6 and 8. Because of the large amount of water consumed the equipment has to be connected to the mains water supply which leads to difficulties in keeping the water temperature constant. The shower is calculated to have a kinetic energy 5.8

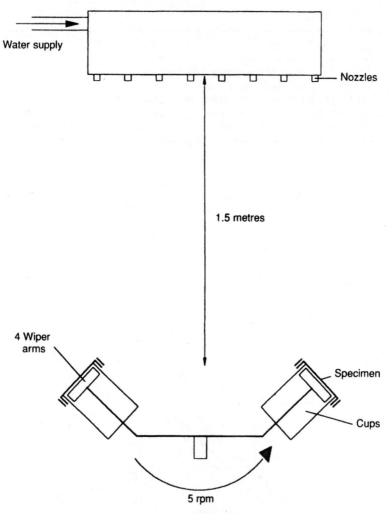

Water supply

Nozzles

1.5 metres

4 Wiper arms

Specimen

Cups

5 rpm

8.15   The Bundesmann shower test.

times that of a cloudburst, 90 times that for heavy rain, 480 times that for moderate rain and 21,000 times that for light rain.

Two fabric parameters are determined from the test:

1   Penetration of water through the fabric: the water collected in the cups is measured to the nearest ml.
2   Absorption of water by the fabric: in order to do this the specimen is weighed before the test and then after the shower. To remove excess water the fabric is shaken ten times using a mechanical shaker and then weighed in an airtight container:

$$\text{Absorption} = \frac{\text{mass of water absorbed}}{\text{original mass}} \times 100\%$$

In each case the mean of four values is calculated.

8.6.3 WIRA shower test

The WIRA shower tester [19] also aims to produce the effect of a rainstorm on a fabric in the laboratory. Like the Bundesmann test it measures water absorption and penetration but it is claimed to have certain advantages over the Bundesmann test. These are:

1   It is not dependent on the flow of water through fine holes in plates or capillary tubes.
2   It uses distilled water and the essential parts are non-metallic so avoiding the corrosion and deposition problems which are associated with tap water.
3   Provision is made for detaching and cleaning the backing material. Penetration is often critically dependent on the surface properties of the backing material.
4   The apparatus is free-standing and is not dependent on elaborate plumbing and electrical fittings.

The method, shown diagrammatically in Fig. 8.16, is based on a patented method of drop propagation which produces a sustained and uniform shower of well-separated drops of water. The method consists of a shallow container having at the base a hydrophobic PTFE (poly (tetrafluoroethene)) plate with large holes covered with a filter paper. Water is run into the container at a fixed rate through a capillary tube from a large funnel.

In the test four samples each one 125 mm × 250 mm are tested two at a time. Firstly the ribbed glass backing plates are cleaned thoroughly by the prescribed method and dried. The filter papers, boxes and measuring cylinders have to be wetted out before the test starts. The specimens are conditioned and then weighed and mounted face side up stretched across· the ribbed glass plate. The valves are opened simultaneously while the stop-watch is started. During the test the time for the first 10 ml of water to collect in the measuring cylinders is noted. The shower should stop after about 7.5 min. After 8.5 min the interceptor is pulled forward and the specimens removed. These are then given ten drops on the mechanical shaker and weighed in a previously weighed airtight container. The water penetration is measured to the nearest 0.5 ml if under 10 ml or to the nearest 1 ml if over 10 ml.

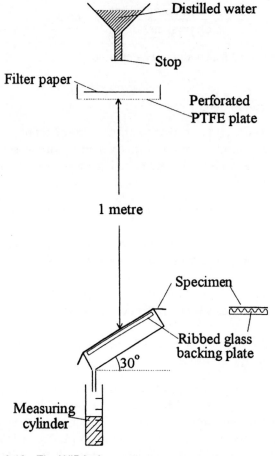

8.16   The WIRA shower test.

Calculate

$$\text{Percentage absorption} = \frac{\text{final mass} - \text{original mass}}{\text{original mass}} \times 100\%$$

Give:

1   the mean absorption %;
2   the mean water penetration in ml;
3   if appropriate mean time for first 10 ml.

This test is not intended for fabrics with marked wicking properties as penetrating water can wick over the edge of the box and so not be recorded. For general rainwear it is suggested that:

- the absorption is not greater than 20%;
- the penetration is not greater than 120 ml;
- the first 10 ml of penetration takes longer than 120 s.

8.6.4 Credit rain simulation tester

This instrument was produced to simulate natural rain in terms of drop size and distribution. The test subjects a fabric to a simulated shower and records the time taken for the water to penetrate to the back of the fabric.

The shower is produced by dropping water through needles onto a mesh cone which splits the drops to give a distribution of drop sizes similar to natural rain. The fabric is mounted over a printed circuit board which is organised so that penetration of water completes a circuit and shows the area where penetration has taken place. An electronic timer is automatically started when the first drop falls and the timing is stopped when contact is made. The time to penetration is shown and an LED (light-emitting diode) display shows where it has occurred.

8.6.5 Hydrostatic head

The hydrostatic head test [20] is primarily intended for closely woven water repellent fabrics. The hydrostatic head supported by a fabric is a measure of the opposition to the passage of water through the fabric. In the test, shown diagrammatically in Fig. 8.17, one face of the fabric is in contact with water which is subjected to a steadily increasing pressure until it penetrates the fabric. The pressure at which the water penetrates the fabric is noted. The current size of specimen used is 100 cm^2.

The circular specimens are clamped between rubber gaskets over a water-filled chamber and pressure is applied to the water forcing it up against the specimen. The pressure of the water is monitored by a water-

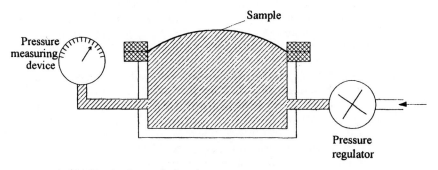

8.17 The hydrostatic head test.

filled manometer which measures the pressure on the specimen in cm of water. The rate of pressure increase of the water is controlled and can be either 10 cm head of water per minute or 60 cm head of water per minute. The results which are obtained at the two rates of increase are not directly comparable.

The pressure is allowed to increase until water appears at three separate places on the surface of the specimen, the pressure being taken at the appearance of the third drop. Five specimens in total are tested and the mean pressure calculated from the results.

The test normally has a maximum of 2 m head of water:

$$1 \, cm \ H_2O = 98.0665 \, Pa$$

A similar test is used for coated fabrics [21] but because the pressures involved are usually higher the requirement for a steady increase in pressure is omitted, otherwise the tests would take a long time. The tests are usually conducted on a pass/fail basis, the pressure being increased to the value laid down in the specification within 1 min from the start of the test.

References

1. Morris J V, 'Performance standards for active wear', *Text Ins Ind*, 1980 **18** 243–245.
2. Farnworth B, 'Mechanisms of heat flow through clothing insulation', *Text Res J*, 1983 **53** 717–725.
3. Fourt L and Hollies N R S, *Clothing Comfort and Function*, Marcel Dekker, New York, 1970.
4. Davies S and Owen P, 'Staying dry and keeping your cool', *Textile Month*, 1989 **Aug** 37.
5. Cooper C, 'Textiles as protection against extreme wintry weather', *Textiles*, 1979 **8** 72.
6. BS 5636 Method of test for the determination of the permeability of fabrics to air.
7. BS 4745 Method for the determination of thermal resistance of textiles.
8. ASTM D 1518 Thermal transmittance of textile materials.
9. BS 7209 Specification for water vapour permeable apparel fabrics.
10. ASTM E 96-80 Standard test methods for water vapor transmission of materials.
11. Gibson P W, 'Factors influencing steady-state heat and water vapour transfer measurements for clothing materials', *Text Res J*, 1993 **63** 749–764.
12. Umbach K H, 'Moisture transport and wear comfort in microfibre fabrics', *Melliand Textilber*, 1993 **74** 174–178.
13. Harnett P R and Mehta P N, 'A survey and comparison of laboratory test methods for measuring wicking', *Text Res J*, 1984 **54** 471.
14. Smith J, 'Comfort in casuals', *Text Horizons*, 1985 **5**(8) 35.
15. Matsudaira M, Watt J D and Carnaby G A, 'Measurement of the surface

prickle of fabrics Part 1: The evaluation of potential objective methods', *J Text Inst*, 1990 **81** 288–299.

16. BS 3449 Testing the resistance of fabrics to water absorption (static immersion test).
17. BS 4554 Method of test for wettability of textile fabrics.
18. BS EN 24920 Textiles. Determination of resistance to surface wetting (spray test) of fabrics.
19. BS 5066 Method of test for the resistance of fabrics to an artificial shower. Method of test for textiles, BS Handbook 11, 1974.
20. BS EN 20811 Textiles. Determination of resistance to water penetration. Hydrostatic pressure test.
21. BS 3424 Testing coated fabrics Part 26 Methods 29A, 29B, 29C and 29D. Methods for determination of resistance to water penetration and surface wetting.

Colour fastness testing

9.1　Introduction

Poor colour fastness in textile products is a major source of customer complaint. The fastness of a colour can vary with the type of dye, the particular shade used, the depth of shade and how well the dyeing process has been carried out. Dyes can also behave differently when in contact with different agents, for instance dyes which may be fast to dry-cleaning may not be fast to washing in water. It is therefore important to test any dyed or printed product for the fastness of the colours that have been used in its decoration. There are a number of agencies that the coloured item may encounter during its lifetime which can cause the colour either to fade or to bleed onto an adjacent uncoloured or light coloured item. These factors vary with the end use for which the product is intended. For instance carpets and upholstery are cleaned in a different way from bed linen and clothing and therefore come into contact with different materials. The agencies that affect coloured materials include light, washing, dry-cleaning, water, perspiration and ironing. There are a large number of colour fastness tests in existence which deal with these agencies and a full list will be found in the British Standard [1]. A further group of tests is connected to processes in manufacturing that the coloured material may undergo after dyeing but before completion of the fabric, processes such as decatising or milling. Despite the fact that the list of colour fastness tests is very long, most of them are conducted along similar lines so that the main differences among the tests are in the agents to which the material is exposed.

Colour fastness is usually assessed separately with respect to:

1　changes in the colour of the specimen being tested, that is colour fading;
2　staining of undyed material which is in contact with the specimen during the test, that is bleeding of colour.

In order to give a more objective result a numerical assessment of each of these effects is made by comparing the changes with two sets of standard grey scales, one for colour change and the other for staining.

1 **Colour change grey scales.** These scales consist of five pairs of grey coloured material numbered from 1 to 5. Number 5 has two identical greys, number 1 grey scale shows the greatest contrast, and numbers 2, 3 and 4 have intermediate contrasts. After appropriate treatment the specimen is compared with the original untreated material and any loss in colour is graded with reference to the grey scale. When there is no change in the colour of a test specimen it would be classified as '5'; if there is a change it is then classified with the number of the scale that shows the same contrast as that between the treated and untreated specimens.

2 **Degree of staining grey scales.** A different set of grey scales is used for measuring staining. Fastness rating 5 is shown by two identical white samples (that is no staining) and rating 1 shows a white and a grey sample. The other numbers show geometrical steps of contrast between white and a series of greys. A piece of untreated, unstained, undyed cloth is compared with the treated sample that has been in contact with the test specimen during the staining test and a numerical assessment of staining is given. A rating of 5 means that there is no difference between the treated and untreated material. If the result is in between any two of the contrasts on the scale, a rating of, for example, 3–4 is given. Sets of grey scales, examples of which are shown in Fig. 9.1, can be supplied by the British Standards Institution.

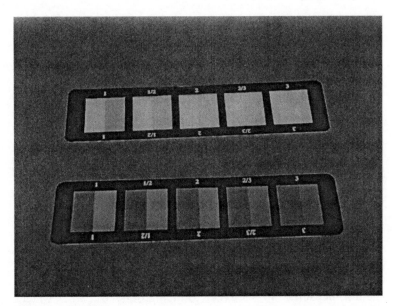

9.1 Grey scales.

Table 9.1 Multifibre strip

Multifibre DW	Multifibre TV
Secondary acetate	Triacetate
Bleached cotton	Bleached cotton
Polyamide	Polyamide
Polyester	Polyester
Acrylic	Acrylic
Wool	Viscose

9.1.1 Sample preparation

Fabric

Fabric is usually tested in the form of a composite specimen (at least 10 cm × 4 cm), made up of the test specimen placed in contact with undyed fabric, usually in the form of multifibre strip, of the same size. Two specific undyed fabrics may be used instead of the multifibre strip; one of these is usually of the same type as the fabric under test and the other is given in the standard. The purpose of the undyed fabric is to measure the staining effect of any dye that has been lost from the test fabric.

There are two types of multifibre adjacent fabric, one with wool which is type DW and one without wool which is type TV. The specification of each is given in Table 9.1.

Fibres

Loose fibres may be tested by compressing them into a pad and sewing it between multifibre strip or undyed cloth.

Yarns

Yarns may be knitted into fabric before testing or sewed between multifibre strip or undyed cloth as above.

9.2   Outline of colour fastness tests

For full experimental procedures see BS EN 20105 [1].

9.2.1 Colour fastness to light

This test measures the resistance to fading of dyed textiles when exposed to daylight. The test is of importance to the dyestuff manufacturer, the dyer and the retailer. Certain end products require a high resistance to fading because of their exposure to light during use, for example: curtains, upholstery, carpets, awnings and coatings. However, many items of apparel also require a degree of light fastness because they are exposed to light when on display, particularly in a shop window.

Light sources

The British Standard allows either daylight or xenon arc light to be used for the test.

Daylight B01

To test the resistance of a material to fading in daylight a sample of it is exposed facing due south (in the northern hemisphere), sloping at an angle from the horizontal which is approximately equal to the test site latitude. The sample is covered with glass and provision is made for it to be ventilated. Together with the specimen under test eight 'standard blue wool dyeings' are exposed. This method gives a true indication of the light fastness of a dyed material but it is slow.

Xenon arc B02

The xenon arc is a much more intense source of light which has a very similar spectral content to that of daylight so that the test is speeded up considerably. Because of the large amount of heat generated by the lamp an efficient heat filter has to be placed between the lamp and specimen and the temperature monitored. This is in addition to a glass filter as above to remove ultra-violet radiation.

Mercury–tungsten fluorescent lamp (MBTF)

This is a source found in certain commercial light fastness testers. It provides a less intense light source than the xenon arc but will still give a faster test result than using daylight. One advantage of it is that the bulbs are cheaper and last longer than do xenon ones. It is claimed to give similar results to daylight [2].

Reference standards

The essence of the test is to expose the sample under test to the light source together with eight blue wool reference standards. The sample and blue standards are partly covered so that some of the material fades and some is left unfaded. A rating is given to the sample which is the number of the reference standard which shows a similar visual contrast between the exposed and unexposed portions as the specimen. This means that the specimen will be given a grade between *one* (poor light fastness) and *eight* (highly resistant to fading). If the result is in between two blue dyeings it is given as 3–4, for example.

There are two sets of blue wool reference standards in use. Those used in Europe are identified by the numerical designation 1 to 8. They range from 1 (very low light fastness) to 8 (very high light fastness) so that each higher numbered reference is approximately twice as fast as the preceding one. The blue wool references used in America are identified by the letter L followed by the numerical designation 2 to 9. The two sets of references are not interchangeable.

Recommended procedure

The sample under test and a set of blue wool reference standards are arranged on a suitable backing as shown in Fig. 9.2. The middle third of the strips is covered with opaque card (A). The assembly is then exposed to light until the specimen just shows a change in shade (4–5 on the grey scale). The number of the standard showing a similar change is noted.

The exposure is continued until the contrast in the specimen is equal to grey scale 4, at which point a second segment of the specimen and standards is covered with another piece of opaque card (B). The exposure is again continued until the contrast between the exposed and unexposed parts of the specimen is equal to grey scale 3, at which point the exposure is terminated.

When the cards are removed the specimen and standards will show two areas that have been exposed for different lengths of time together with an unexposed area. The specimen is given the rating of the standard which shows similar changes. If the exposed areas have different ratings then the overall rating is the mean of the two ratings.

If the grade given to the specimen is 4 or higher then the initial assessment at a contrast 4–5 on the grey scale becomes significant as some colours can fade at a faster rate on initial exposure to light. If this initial grade is different from the main grade it is included after the main grade in brackets, for example 6(3).

Photochromism

Some dyes change colour rapidly on exposure to a strong light but on being put in the dark the original colour returns to a greater or less extent. This

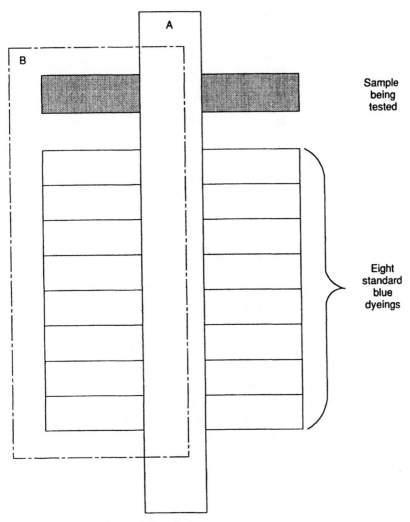

9.2   Sample for light fastness test.

is known as photochromism. To avoid any error due to this effect samples should be conditioned for 24 h in the dark before assessment.

Change in hue

Certain dyed materials change hue on prolonged exposure to light, for instance a yellow may become brown or a purple may become blue. The rating for colour fastness concerns only change in contrast of the dyed mate-

Table 9.2 Wash test conditions

Test	Liquor	Temp. (°C)	Time (min)	Reproduces action of
C01	0.5% soap	40	30	Hand washing
C02.	0.5% soap	50	45	Repeated hand washing
C03	0.5% soap	60	30	Medium cellulosic wash
	0.2% soda ash			Severe wool wash
C04	0.5% soap	95	30	Severe cellulosic wash
	0.2% soda ash			
C05	0.5% soap	95	240	Very severe cellulosic
	0.2% soda ash			Wash
C06	4 g/l reference	Various	Various	Domestic laundering
	detergent +			
	perborate			

rial which in such cases may not have altered. In such cases the change in hue is included as part of the rating depending on the blue dyeing which has changed at the same time. A rating of '5 bluer' would be used for a sample which changed from green to blue at the same rate as the reference sample 5.

Effect of humidity

The fastness of dyes is affected by both temperature and humidity. Increases in temperature and humidity both increase the rate of fading of dyes. The effect of humidity is much greater than that of temperature. However the humidity that affects colour fastness is the humidity at the surface of the specimen which can be modified by the surface temperature of the specimen. The specimen temperature is raised by the absorption of radiation and this has the effect of lowering the relative humidity in the immediate region of the specimen. In order to monitor the effective humidity a standard fabric of red azoic dyed cotton can be used, exposed at the same time as the test specimen. The light fastness at different relative humidities of this fabric is known.

9.2.2 Colour fastness to other agents

Fastness to washing C01–C06

A composite specimen is agitated in a wash-wheel using one of the sets of conditions shown in Table 9.2. The sample is then dried and assessed for colour loss and the adjacent fabric is assessed for staining.

A wash-wheel consists of a number of closed stainless steel containers rotating in a water bath at 40 revolutions per minute. The sample and appropriate liquor are sealed inside the container and the water bath is heated to the desired temperature.

Dry-cleaning D01

The sample to be tested is enclosed in an undyed cotton bag with 12 steel discs. The bag is then agitated in a wash-wheel for 30 min in perchloroethylene at 30 °C. The specimen is then removed, dried and then assessed for colour change. The colour of the remaining solvent is also compared with that of unused solvent to detect any staining.

Colour fastness to water E01

In this test composite specimens are wetted out in distilled water at room temperature. They are then placed in a perspirometer in an oven at 37 °C for 4 h, removed and dried. They are then assessed for colour change of the test fabric and staining of the adjacent fabric.

A perspirometer, an example of which is shown in Fig. 9.3, consists of a stainless steel frame constructed to hold a number of glass or acrylic plates each measuring 60 mm × 115 mm. The samples of size 40 mm × 100 mm are each placed separately between a pair of these plates in order to keep them moist. A mass of 5 kg is then placed on top of the apparatus so as to apply a pressure of 12.5 kPa to each specimen. The perspirometer

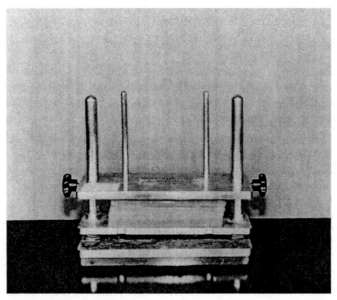

9.3 Perspirometer.

is so constructed that when the mass is removed the specimens remain under pressure.

Seawater E02

The composite specimen is wetted out in sodium chloride solution (30 g/l) and placed in a perspirometer in an oven for 4 h at 37 °C. It is then separated, dried and assessed for colour change of the test fabric and staining of the adjacent fabric.

Chlorinated water (swimming-pool water) E03

The specimen is agitated in a wash-wheel for 1 h at 27 °C in a weak solution of sodium hypochlorite (either 20 mg/l active chlorine for towels or 100 mg/l for swimwear), dried and assessed for colour change.

Colour fastness to perspiration E04

This method is intended for the determination of the resistance of the colour of textiles of all kinds and in all forms to perspiration.

Composite specimens are treated in solutions containing histidine, drained and placed in a Perspirometer or equivalent apparatus. The specimen and the undyed cloths are dried separately. The change in colour of the specimens and the staining of the undyed cloths are assessed with standard grey scales.

Solutions

1   Alkaline solution, *freshly prepared*, containing 0.5 g histidine monohydrochloride monohydrate, 5 g sodium chloride and 2.5 g disodium hydrogen orthophosphate ($Na_2HPO_4 \cdot 2H_2O$) *per litre*, brought to pH 8 with 0.1 N sodium hydroxide.
2   Acid solution, *freshly prepared*, containing 0.5 g histidine monohydrochloride monohydrate, 5 g sodium chloride and 2.2 g sodium dihydrogen orthophosphate ($NaH_2PO_4 \cdot 2H_2O$) *per litre*, brought to pH 5.5 with 0.1 N sodium hydroxide.

Procedure

Thoroughly wet one composite specimen in the solution at pH 8 (solution 1 above) at a liquor ratio of 50:1 and allow it to remain in the solution for 30 min at room temperature. Wipe excess liquid off the specimen between two glass rods and place the specimen between two plates of the perspirometer under a pressure of 12.5 kPa. Repeat with the other composite specimen in the acid solution using a separate perspirometer.

Place the perspirometers in an oven at 37 °C for 4 h.
Remove specimens, open out and allow to dry.

Both specimens are then assessed for colour change of the test fabric and staining of the adjacent fabric.

Fastness to acid spotting E05

The specimen is spotted with either acetic, sulphuric or tartaric acid which is rubbed into the fabric with a glass rod to form a 20 mm diameter spot. The specimen is then dried and compared with the grey scale for colour change.

Alkali spotting E06

This is carried out as the above test for acid spotting but using a 10% sodium carbonate solution.

Potting E09

The composite specimen is rolled around a glass rod, tied with thread, boiled for 1 h in water; the dyed specimen is then separated from the undyed cloth, dried and assessed for colour change of the test fabric and staining of the adjacent fabric.

Decatising E10

The specimen is wrapped around a perforated steam cylinder for 15 min (mild or severe treatment as required) together with a test control specimen. The control specimen is one that has been dyed to a given formula and that should change colour to a known degree if the procedure has been carried out correctly. The specimen is then dried and assessed for colour change.

Steaming E11

This test is carried out to determine the effect that the steaming stage in printing has on the colours. In the test steam is passed at atmospheric pressure through the composite specimen for 30 min, the adjacent fabric is then assessed for staining.

Colour fastness to nitrogen oxides G01

This method determines the resistance to fading of dyed fabrics by gases produced during the combustion of gas, coal, oil, etc. In the test the sample is exposed to nitric oxide inside a special test chamber. At the same time a control sample is exposed and the test is allowed to proceed until this fades to a given extent. The sample is then assessed for fading.

Colour fastness to burnt gas fumes G02

This method determines the resistance to fading of dyed fabrics to the nitrogen oxides produced during the combustion of butane. In the test the sample and control fabrics are exposed to the fumes of a butane burner in a test chamber. The test is continued until the control fabric has faded to a given extent. The sample is then assessed for fading.

Bleaching with sodium hypochlorite N01

In this test the sample is wetted out and placed in sodium hypochlorite solution of a standard strength (2.0 g/l available chlorine pH 11) and liquor ratio for 1 h; it is rinsed, immersed in hydrogen peroxide solution (30%) for 10 min, rinsed again, dried and assessed for change in colour.

Bleaching with peroxide N02

A composite specimen is placed in a test-tube with standard bleaching solution (composition, temperature and time vary for different fibres, e.g. wool in hydrogen peroxide solution for 2 h). It is then removed, rinsed, squeezed out, opened out, dried, and then assessed for colour change of the test fabric and staining of the adjacent fabric.

Milling E12

The composite specimen is treated in a wash-wheel with additional agitation (ball bearings) using a soap solution for alkaline milling. The treatment is carried out for a fixed time, the specimen is then rinsed, separated and dried, and assessed for colour change of the test fabric and staining of the adjacent fabric.

Carbonising X02

The composite specimen is immersed in sulphuric acid solution (50 g/l) for 15 min at room temperature; removed, squeezed, dried, and then baked for 15 min at 105 °C. The colour change is then assessed. A control specimen is treated at the same time to ensure that the test has been correctly performed. This is a fabric that has been dyed to a given formula and that should change colour to a known degree if the procedure has been carried out correctly.

Cross-dyeing (wool) X07

In this test composite specimens are treated in different dye baths containing all the chemicals except the dye (e.g. sodium sulphate and acetic acid for wool), rinsed and dried. The composite specimens are then separated and assessed for colour change of the test fabric and staining of the adjacent fabric.

Pressing X11

Fastness to pressing can be carried out under three different conditions, dry, damp or wet:

1 Dry – the specimen is placed on dry undyed cotton adjacent fabric and pressed for 15 s with a heated press (a domestic iron may be used, cotton 200 ± 2 °C; wool 150 ± 2 °C). It is then assessed immediately and after a 4 h interval for colour change and the adjacent fabric for staining.
2 Damp – the dry specimen is placed on dry cotton adjacent fabric, covered with wet cotton adjacent fabric and pressed for 15 s. It is then assessed as before.
3 Wet – the specimen is wetted and placed on dry cotton adjacent fabric and then covered with wet cotton adjacent fabric and pressed for 15 s. It is then assessed as before.

Rubbing X12

In this test the dyed specimens are rubbed 10 times using a Crockmeter which has a weighted finger covered with piece of undyed cotton cloth 5 cm × 5 cm. For wet rubbing the cotton cloth is wetted out before being rubbed on the dyed sample. The cotton rubbing cloth is then examined for dye which may have been removed and assessed using the grey scales for staining.

References

1. BS EN 20105 Textiles. Tests for colour fastness.
2. Hindson W R and Southwell G, 'The mercury–tungsten fluorescent lamp for the fading assessment of textiles', *Tex Inst Ind*, 1974 **12** 42–45.

10

Objective evaluation of fabric handle

10.1 Handle

Fabric end uses can be roughly divided into industrial, household and apparel. Fabrics for industrial uses can be chosen on straightforward performance characteristics such as tensile strength, extension and resistance to environmental attack. However, fabrics intended for clothing have less emphasis placed on their technical specification and more on their appearance and handling characteristics such as lustre, smoothness or roughness, stiffness or limpness and draping qualities. Handling the fabric is one of the ways of assessing certain of these properties. 'Handle', the term given to properties assessed by touch or feel, depends upon subjective assessment of the fabrics by a person. Terms such as smooth, rough, stiff or limp depend strongly on the type of fabric being assessed, for instance the smoothness of a worsted suiting is different in nature from that of a cotton sateen. Because of the subjective nature of these properties attempts have been made over the years to devise objective tests to measure some or all of the factors that go to make up handle. Fabric stiffness and drape were some of the earliest [1] properties to be measured objectively.

10.1.1 Bending length

A form of the cantilever stiffness test is often used as a measure of a fabric's stiffness as it is an easy test to carry out. In the test a horizontal strip of fabric is clamped at one end and the rest of the strip allowed to hang under its own weight. This is shown diagrammatically in Fig. 10.1.

The relationship among the length of the overhanging strip, the angle that it bends to and the flexural rigidity, G, of the fabric is a complex one which was solved empirically by Peirce [1] to give the formula:

$$G = ML^3 \left(\frac{\cos \frac{1}{2}\theta}{8 \tan \theta} \right)$$

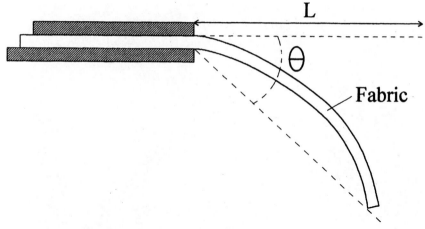

10.1   Bending length.

where L = length of fabric projecting,
     θ = angle fabric bends to,
     M = fabric mass per unit area.

From this relationship Peirce defined a quantity known as the bending length as being equal to the length of a rectangular strip of material which will bend under its own mass to an angle of 7.1° [1]. The bending length is dependent on the weight of the fabric and is therefore an important component of the drape of a fabric when it is hanging under its own weight. However, when a fabric is handled by the fingers the property relating to stiffness that is sensed, in this situation, is the flexural rigidity which is a measure of stiffness independent of the fabric weight.

The bending length is related to the angle that the fabric makes to the horizontal by the following relation:

$$ C = L \left(\frac{\cos \frac{1}{2}\theta}{8 \tan \theta} \right)^{1/3} $$

where C = bending length.

When the tip of the specimen reaches a plane inclined at 41.5° below the horizontal the overhanging length is then twice the bending length. This angle is used in the Shirley apparatus (Fig. 10.2) to increase the sensitivity of the length measurement and the slide on this instrument is directly calibrated in centimetres.

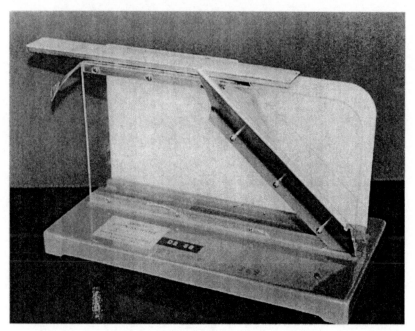

10.2   The Shirley stiffness test.

10.1.2 Shirley stiffness test

This test [2] measures the bending stiffness of a fabric by allowing a narrow strip of the fabric to bend to a fixed angle under its own weight. The length of the fabric required to bend to this angle is measured and is known as the bending length.

The test specimens are each 25 mm wide and 200 mm long; three are cut parallel to the warp and three parallel to the weft so that no two warp specimens contain the same warp threads, and no two weft specimens contain the same weft threads. The specimens should not be creased and those that tend to twist should be flattened.

Before the test the specimens are preconditioned for 4 h (50 °C >10% RH) and then conditioned for 24 h. If a specimen is found to be twisted its mid-point should be aligned with the two index lines. Four readings are taken from each specimen, one face up and one face down on the first end, and then the same for the second end.

The mean bending length for warp and weft is calculated. The higher the bending length, the stiffer is the fabric.

Flexural rigidity

The flexural rigidity is the ratio of the small change in bending moment per unit width of the material to the corresponding small change in curvature:

$$\text{Flexural rigidity } G = M \times C^3 \times 9.807 \times 10^{-6} \mu N\,m$$

where C = bending length (mm),

   M = fabric mass per unit area (g/m^2).

Bending modulus

The stiffness of a fabric in bending is very dependent on its thickness, the thicker the fabric, the stiffer it is if all other factors remain the same. The bending modulus is independent of the dimensions of the strip tested so that by analogy with solid materials it is a measure of 'intrinsic stiffness'.

$$\text{Bending modulus} = \frac{12 \times G \times 10^3}{T^3} N/m^2$$

where T = fabric thickness (mm).

10.1.3 Hanging loop method

Fabrics that are too limp to give a satisfactory result by the cantilever method may have their stiffness measured by forming them into a loop and allowing it to hang under its own weight. A strip of fabric of length L has its two ends clamped together to form a loop. The undistorted length of the loop l_0, from the grip to the lowest point, has been calculated [1] for three different loop shapes: the ring, pear and heart shapes as shown in Fig. 10.3. If the actual length l of the loop hanging under its own weight is measured the stiffness can be calculated from the difference between the calculated and measured lengths $d = l - l_0$:

Ring loop: $l_0 = 0.3183L$ $\theta = 157° \dfrac{d}{l_0}$

Bending length $C = L0.133 f_2(\theta)$

Pear loop: $l_0 = 0.4243L$ $\theta = 504.5° \dfrac{d}{l_0}$

Bending length $C = L0.133 f_2(\theta)/\cos 0.87\theta$

Heart loop: $l_0 = 0.1337L$ $\theta = 32.85° \dfrac{d}{l_0}$

Bending length $C = l_0 f_2(\theta)$ $f_2(\theta) = \left(\dfrac{\cos\theta}{\tan\theta}\right)^{1/3}$

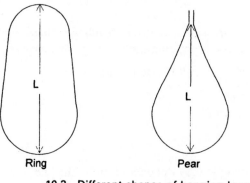
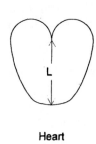

Ring   Pear   Heart

10.3   Different shapes of hanging loops.

10.1.4 Drape

Drape is the term used to describe the way a fabric hangs under its own weight. It has an important bearing on how good a garment looks in use. The draping qualities required from a fabric will differ completely depending on its end use, therefore a given value for drape cannot be classified as either good or bad. Knitted fabrics are relatively floppy and garments made from them will tend to follow the body contours. Woven fabrics are relatively stiff when compared with knitted fabrics so that they are used in tailored clothing where the fabric hangs away from the body and disguises its contours. Measurement of a fabric's drape is meant to assess its ability to do this and also its ability to hang in graceful curves.

Cusick drape test

In the drape test [3] the specimen deforms with multi-directional curvature and consequently the results are dependent to a certain amount upon the shear properties of the fabric. The results are mainly dependent, however, on the bending stiffness of the fabric.

   In the test a circular specimen is held concentrically between two smaller horizontal discs and is allowed to drape into folds under its own weight. A light is shone from underneath the specimen as shown in Fig. 10.4 and the shadow that the fabric casts, shown in Fig. 10.5, is traced onto an annular piece of paper the same size as the unsupported part of the fabric specimen. The stiffer a fabric is, the larger is the area of its shadow compared with the unsupported area of the fabric. To measure the areas involved, the whole paper ring is weighed and then the shadow part of the ring is cut away and weighed. The paper is assumed to have constant mass per unit area so that the measured mass is proportional to area. The drape coefficient can then be calculated using the following equation:

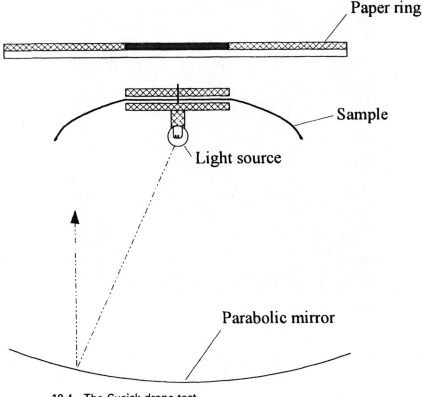

10.4   The Cusick drape test.

$$\text{Drape coefficient} = \frac{\text{mass of shaded area}}{\text{total mass of paper ring}} \times 100\%$$

The higher the drape coefficient the stiffer is the fabric.

At least two specimens should be used, the fabric being tested both ways up so that a total of six measurements are made on the same specimen.

There are three diameters of specimen that can be used:

- A 24 cm for limp fabrics; drape coefficient below 30% with the 30 cm sample;
- B 30 cm for medium fabrics;
- C 36 cm for stiff fabrics; drape coefficient above 85% with the 30 cm sample.

It is intended that a fabric should be tested initially with a 30 cm size specimen in order to see which of the above categories it falls into.

When test specimens of different diameter are used, the drape coefficients measured from them are not directly comparable with one

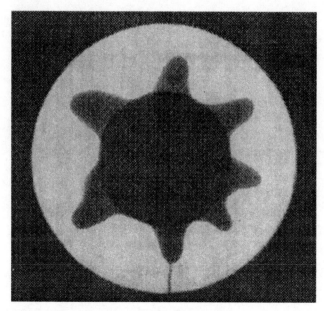

10.5 Drape test, top view of draped fabric.

another. Figure 10.6 shows a drape tester fitted with a video camera and computer for instantaneous measurement of the drape coefficient.

10.1.5 Crease recovery

Creasing of a fabric during wear is not a change in appearance that is generally desired. The ability of a fabric to resist creasing is in the first instance dependent on the type of fibre used in its construction. Some fibre types such as wool and cultivated silk have a good resistance to creasing whereas cellulosic materials such as cotton, viscose and linen have a very poor resistance to creasing. Many fabrics have resin finishes applied during production in order to improve their crease resistance. This test was originally developed to test the efficiency of such finishes.

The essence of the test [4] is that a small fabric specimen is folded in two and placed under a load for a given length of time to form a crease and it is then allowed to recover for a further length of time and the angle of the crease that remains is measured.

The magnitude of this crease recovery angle is an indication of the ability of a fabric to recover from accidental creasing. Some types of fabrics, owing to limpness, thickness and tendency to curl, give rise to ill-defined crease recovery angles and therefore imprecise measurements. Many wool and wool mix fabrics come under this heading, therefore a different test using smaller specimens is used in this case.

10.6   Drape test.

The test can be carried out in two atmospheres, either the standard one or at 90% RH and 35 °C.

Twenty rectangular specimens are tested, each measuring 40 mm × 15 mm, half of the specimens cut parallel to the warp and half parallel to the weft.

In the test the specimens are folded in two, the ends being held by tweezers. Half the specimens are folded face to face and half of them back to back. The specimens are then placed under a 10 N load for 5 min. They are then transferred immediately to the holder of the measuring instrument and one leg of the specimen is inserted as far as the back stop. The instrument is adjusted continuously to keep the free limb of the specimen vertical as shown in Fig. 10.7. The crease recovery angle is measured, by reading the scale when the free limb is vertical, 5 min after the removal of the load.

The following mean values are calculated

    warp face to face   weft face to face
    warp back to back   weft back to back

When a fabric is creased the resulting deformation has two components: one is the displacement of fibres and yarns relative to one another and the second is the stretching of the fibres on the outside of the curve. The

10.7 Crease recovery.

relative importance of these two mechanisms depends on the radius of the curve that the fabric is bent into. The smaller the radius of curvature, the more likely it is that the fibres are actually stretched rather than the curvature being accommodated by fibre displacement.

The unaided recovery of the fabric from creasing depends on the elastic recovery of the fibres, in particular whether the stored elastic energy is sufficient to overcome the friction that resists the movement of the yarns and fibres. Crease recovery in both resin treated and untreated cotton fabrics has been found [5] to increase with decreasing curvature but tending to the same limiting value at less than 100% recovery. This is thought to be because the crease recovery at low curvatures is governed by the frictional effects associated with fibre movement and at high curvatures by the elastic response of the fibre. The effect of the resin treatment is to improve the fibres' elastic recovery but it does not markedly affect the internal friction of the fabric which is dependent on structural factors such as tightness of weave.

The elastic recovery of the fibres is dependent on the time-related effects, such as stress relaxation, detailed in section 5.3.4. Hilyard et al. [5] have shown that the recovery from creasing of a fabric is a function of both the time the crease is maintained and the time allowed for recovery. There is an initial rapid recovery which takes place after removal of the restraint followed by a much slower rate of recovery which decreases with time.

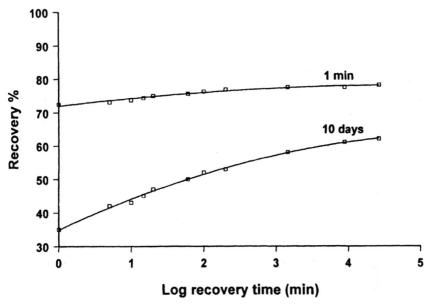

10.8   Improvement in crease recovery as a function of recovery time
for two different loading times: 1 min and 10 days. From [5].

The recovery with time of two identical samples, one creased for 1 min and
the second creased for 10 days, is shown in Fig. 10.8 which is based on data
from [5].

10.1.6 Fabric thickness

It might be expected that the thickness of a fabric is one of its basic properties giving information on its warmth, heaviness or stiffness in use. In
practice thickness measurements are rarely used as they are very sensitive
to the pressure used in the measurement. Instead fabric weight per unit area
is used commercially as an indicator of thickness.

Besides fibres a fabric encloses a large amount of air, which among other
things, is responsible for its good thermal insulation properties. When a
fabric is compressed, the space between the fibres is decreased until they
eventually come into contact with one another. Three stages in the deformation of a fabric have been identified [6]. Firstly the individual fibres protruding from the surface are compressed. The resistance to compression in
this region comes from the bending of the fibres. Secondly contact is made
with the surface of the yarn, at which point the inter-yarn and/or inter-fibre
friction provides the resistance to compression until the fibres are all in
contact with one another. In the third stage the resistance is provided by
the lateral compression of the fibres themselves.

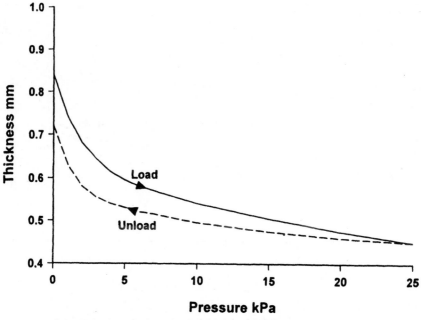

10.9   The change in thickness with pressure.

Matsudaira and Qin [6] consider that in the first and third stages of compression elastic deformation is taking place, whereas in the second stage it is frictional forces that have to be overcome both in compression and also in the subsequent recovery. The forces, which cause the fabric to regain most of its original thickness after compression, come from the elastic recovery of the fibres from bending and lateral compression.

Figure 10.9 shows the change in thickness with pressure for a soft fabric together with the recovery in thickness as the pressure is removed. The steep initial slope of the curve makes it very difficult to measure thickness with any accuracy as a small change in pressure in this region causes a large change in measured thickness. Thickness at zero pressure always has to be obtained by extrapolation of the curve, as a positive pressure is needed to bring any measuring instrument into contact with the fabric surface.

The hysteresis between the loading and unloading curves is due to the internal friction of the fabric. The difference in thickness at a given low pressure between the loading and unloading cycles can be used as a measure of resilience. There is, however, a time element involved as the fabric thickness can recover slowly with time after being compressed.

Matsudaira and Qin [6] consider that it is the second stage of the loading curve that contains information about the handle of the fabric. The greater is the radius of curvature of the transition between the first and third stages, the softer is the fabric in compression.

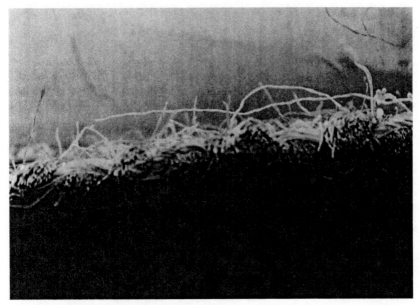

10.10   Surface hairs ×30.

Optical methods have been put forward as a way of measuring fabric thickness as they do not require any physical contact with the fabric surface. However, the problem of such methods is that of defining precisely where the surface starts. Most fabrics have loose fibres that protrude some way above the surface as shown in Fig. 10.10 and the density of these increases as the surface is approached. In brushed and raised fabrics these surface fibres are an important part of the fabric thickness as is shown in Fig. 10.11. Defining the critical point where the fibres end and the surface proper begins therefore relies on the judgement of the operator, unlike measurements involving surface contact where an agreed pressure can be used. A fabric made from continuous filament yarns is shown in Fig. 10.12 for comparison.

10.1.7 Shear

The behaviour of a fabric when it is subjected to shearing forces is one of the factors that determines how it will perform when subjected to a wide variety of complex deformations during use. The ability of a fabric to deform by shearing differentiates it from other thin sheet materials such as paper or plastic film. It is this property that enables it to undergo more complex deformations than two-dimensional bending and so conform to the contours of the body in clothing applications.

10.11   A fabric with a raised finish ×16.

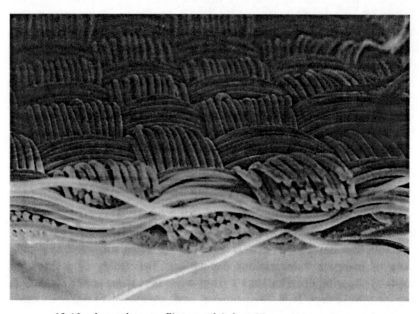

10.12   A continuous filament fabric ×60.

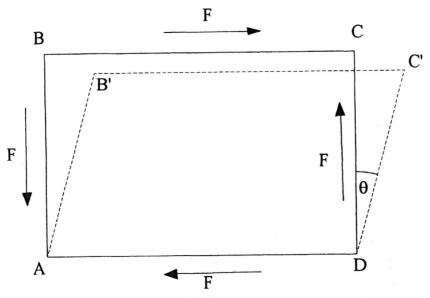

10.13   The distortion due to shear.

The behaviour in shear of textile materials cannot be analysed by the means which are applied to homogeneous materials; however, a simplified analysis of the shear of fabrics has been developed by workers in this area [7]. The basic situation is shown in Fig. 10.13: a rectangular element of material ABCD is subjected to pairs of equal and opposite stresses F which are acting parallel to the side of the element. In the case of simple shear it is assumed that the element deforms to a position shown by AB′C′D in such a way that its area remains constant. The shear strain is defined as the tangent of the change in angle between the side of the element θ. For elastic materials there is a linear relationship between shear stress F and the shear strain $\tan \theta$:

$$F = G \tan \theta$$

where G is the shear modulus.

However, the shear deformation that is found in fabrics is not in general a simple shear at constant area nor does it confrm to any other simple theoretical model such as the length of the sides of the original rectangle remaining constant.

The forces acting when a material is in shear as shown in Fig. 10.13 are equivalent to an extension acting along the diagonal AC and a corresponding compression which acts along the diagonal BD. In practice these forces give rise to problems in measuring shear properties because fabrics subjected to compressive forces in the plane of the material will

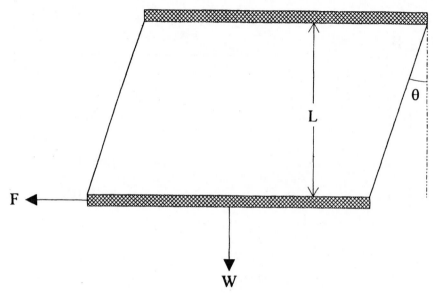

10.14   Diagram of shear test method.

tend to buckle at very low values. It is possible to delay the onset of buck-ling by putting the fabric under tension so as to oppose the compressive force.

A method used [8] to measure shear is shown diagrammatically in a simplified form in Fig. 10.14. In the method the fabric is held rigidly by clamps at the top and bottom. A vertical force W is applied to the fabric by using a weighted bottom clamp in order to delay the onset of buckling. The horizontal force F which is required to move the bottom clamp laterally is measured together with the shear angle θ. However, in this experimental configuration the applied force F is not equal to the shearing force as a quantity $W\tan\theta$ has to be subtracted from the applied force. This factor arises because as the clamp is displaced laterally it is also raised vertically so that an extra force of $W\tan\theta$ has to be supplied in order to do this. Therefore:

Effective shear force $= F - W\tan\theta$

The force is usually expressed as force per unit length.

Treloar has shown [8] that the errors associated with the onset of wrin-kling can be reduced by the use of a narrow specimen with a reduced dis-tance between the clamps instead of a square one. A height:width ratio of 1:10 is considered to be the limit for practical measurements.

More refined versions of this apparatus have been designed [9] to fit directly onto standard tensile testing machines so that shearing can take

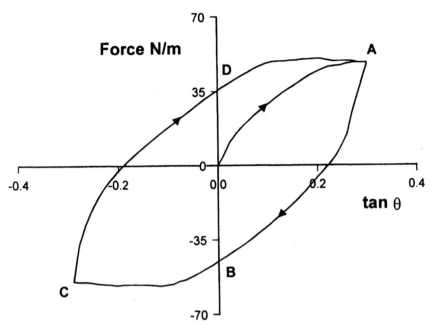

10.15   An example of shear.

place in each direction. With such apparatus a full shear stress–strain curve can be plotted over one full cycle, a specimen of which is shown in Fig. 10.15. Initially the line from the origin is followed to A, at which point the load is reversed and the line then goes through B to C. At this point the sample has been sheared to the same angle in the opposite direction, the load is again reversed and the sample is taken through a further half cycle back to A. The path through ABCD will then be followed on any subsequent shearing cycle. It can be seen from this example that hysteresis occurs when the direction of shear is reversed. This is due to the fact that when a fabric is sheared, most of the force expended is used in overcoming the frictional forces that exist at the intersection of warp and weft. These frictional forces always oppose the applied shearing force whichever direction it is applied. Figure 10.16 shows a fabric that has a lower shear stiffness than that shown in Fig. 10.15.

A number of ways of quantifying shear behaviour have been proposed [9, 10]; these include:

1   The initial shear modulus given by the slope of the curve at the origin.
2   The shear modulus at zero shear angle given by the slope at points B and D (Fig. 10.15).
3   The hysteresis at zero shear angle given by the length BD in the diagram.

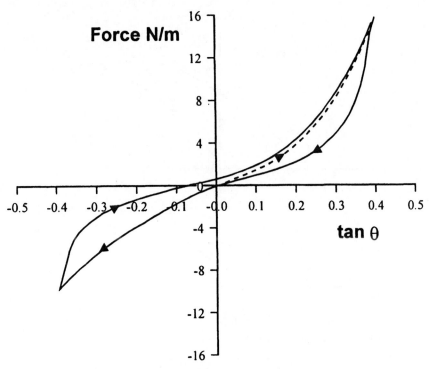

10.16   A further example of shear.

10.1.8 Bias extension

Kilby [11] has derived a formula which gives the Young's modulus for a fabric in directions which are at an angle to the warp direction. From this work Leaf and Sheta [12] have shown that if a fabric is extended in a direction that makes an angle of 45° with the warp threads the Young's modulus in that direction (bias direction) E_{45} is connected to the shear modulus G by the following equation:

$$\frac{1}{G} = \frac{4}{E_{45}} - \frac{1 - \sigma_2}{E_1} - \frac{1 - \sigma_1}{E_2}$$

where E_1 and E_2 are the Young's moduli in the warp and weft directions and σ_1 and σ_2 are the fabric Poisson's ratios.

Generally the modulus in the bias direction is much lower than in the warp and weft directions so that the modulus in the bias direction is determined predominantly by the shear modulus. If the warp and weft moduli are much greater than the bias modulus it may be possible to simplify the expression to give:

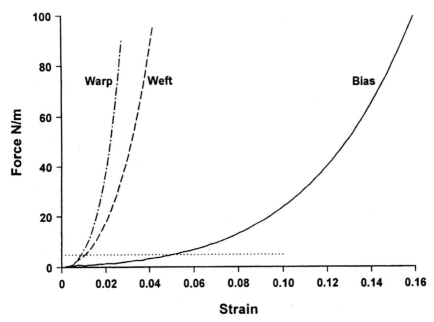

10.17   Force extension curve of warp, weft and bias directions.

$$E_{45} \approx 4G$$

However, Leaf and Sheta [12] demonstrated that when the measured values of warp and weft moduli and Poisson's ratios are taken into account they can significantly alter the calculated shear modulus for some fabrics. Figure 10.17 shows the initial force extension curves for the warp, weft and bias directions of the fabric sample whose shear deformation is shown in Fig. 10.16. The horizontal dotted line corresponds to a force of 5 gf/cm, which is the force which is used in the bias extension measurement for the FAST system (see below).

Spivak and Treloar [13] also analysed the bias extension of fabrics but made the assumption of inextensible warp and weft yarns so that a fabric acts like a trellis pivoted at the thread intersections as shown diagrammatically in Fig. 10.18. They calculated that the shear strain in simple shear is equivalent to:

$$\tan\theta \approx 2e + e^2$$

where e is the bias extension.

For infinitesimal strains this reduces to

$$\text{Shear strain} = 2e$$

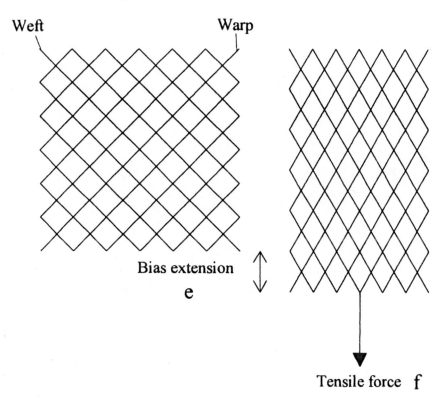

Weft

Warp

Bias extension

e

Tensile force f

10.18   The distortion due to bias extension.

The corresponding shear stress for infinitesimal strains is:

$$F - W\tan\theta \approx \frac{f}{2}$$

where f is the applied tensile force and F and W have the same meanings as in simple shear above. The forces are usually expressed in terms of force per unit length.

   Spivak and Treloar [13] found experimentally that there were inconsistencies between the fabric properties measured in simple shear and by bias extension. These were considered to be due to a number of factors including: the geometry of the test specimen, the assumption in the analysis that the threads were inextensible and the variation that takes place in the normal stress W during bias extension.

10.1.9 Formability

Lindberg *et al.* [14] investigated the fabric properties that are specifically required in garment construction. Among other properties they identified the need for a fabric to be able to be compressed in the plane of the fabric without buckling. For instance at the cuff or collar of a garment the fabric is turned over on itself which means that the inner layer of fabric has to conform to a smaller radius of curvature than the outer layer. In order to do this the outer layer has to stretch and the inner layer has to contract. If the fabric is unable to accommodate this change in length the inner layer will pucker. The ability to deform in this manner was given the title of formability and it is a measure of the amount of compression that a fabric can undergo before it buckles.

The measurement of formability is derived from the bending stiffness of the fabric and its modulus of compression. The compression modulus cannot be measured directly as the fabric quickly buckles. It is, however, derived from the extension modulus by assuming that at small strains, around zero on the force extension curve, the slope of the curve is the same at positive and negative stresses as shown in Fig. 10.19. From this assumption:

$$B = CP$$

where B = compression,
P = force,
C = compressibility, that is the slope of the force extension curve.
The force required to buckle a sample of fabric of length l is given by:

$$P = k\frac{b}{l^2}$$

where k = constant,
b = bending rigidity.
Substituting for P in this equation, the amount that a fabric of length l can be compressed before it buckles is then given by:

$$B = k\frac{Cb}{l^2}$$

Within the limits of this equation the product Cb is a specific property of the fabric which determines how much compression it can undergo before buckling. Lindberg terms this product the compression formability F_c. For a given fabric the formability will vary with direction as both the modulus and bending stiffness vary with direction.

10.1.10 Fabric friction

Fabric friction is subject to the same rules as yarn friction which were outlined in Chapter 4. Two main ways are generally used to measure fabric fric-

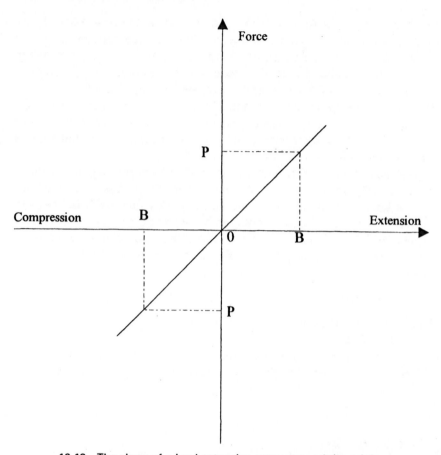

10.19   The slope of a load extension curve around the origin.

tion. One of these methods is shown diagrammatically in Fig. 10.20. In this method a block of mass m is pulled over a flat rigid surface which is covered with the fabric being tested. The line connected to the block is led around a frictionless pulley and connected to an appropriate load cell in a tensile testing machine. This can measure the force F required both to start the block moving and also to keep it moving, thus providing the static and dynamic coefficients of friction from the relation:

$$\text{Coefficient of friction } \mu = \frac{F}{mg}$$

If a chart recorder is available a trace of frictional force can be obtained which contains further information on the frictional properties. Figure 10.21 shows this form of friction measurement as an attachment for a standard tensile tester.

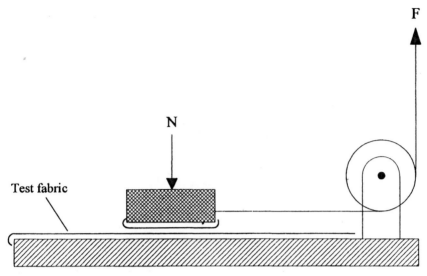

10.20   Friction test.

10.21   Friction apparatus.

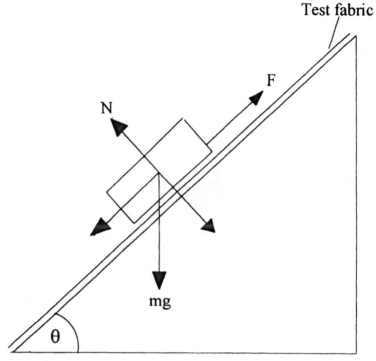

10.22   Friction inclined plane.

The coefficient of friction which is measured is specific for the two materials in contact so that the choice of material for the block is important. The block used may be a solid construction of a known material such as wood or steel or it may be covered in fabric. In the case of a fabric covering the choice is between a standard fabric which is used for all friction tests or a portion of the fabric which is being tested. The use of the same fabric for both surfaces is the preferred option as it removes any problems of standardisation of the block surface. A factor that can affect fabric friction measurements is the presence on the fabric of finishes such as softeners which reduce the fabric friction. These can easily be transferred from the fabric to the block so that it needs to be cleaned or covered with a fresh piece of fabric before every test.

The second method used for measuring fabric friction is the inclined plane. This is shown diagrammatically in Fig. 10.22 which shows a block of mass m initially resting on an inclined plane covered with the fabric to be tested. The apparatus is arranged so that the angle of the plane θ can be continuously adjusted until the block just begins to slide. At this point the frictional force F is equal to the component of the mass of the block parallel to the inclined plane:

$$F = mg \sin \theta$$

The normal reaction N is equal to the component of the mass perpendicular to the inclined plane

$$N = mg \cos \theta$$

As the coefficient of friction $\mu = F/N$. Therefore:

$$\mu = \frac{mg \sin \theta}{mg \cos \theta}$$

$$= \tan \theta$$

This procedure as described measures the coefficient of static friction. The coefficient of dynamic friction may be measured by giving the block an initial impetus and determining the angle at which motion just continues.

With textile materials the coefficient of friction is found to be dependent on the normal force N instead of being independent of it as would be expected from Amonton's laws. The relationship:

$$F = aN^n$$

where a is a coefficient (equal to μ only when $n = 1$) and n is the friction index which can vary between 0.67 and 1.0, has been found [15] to fit the experimental data more closely.

10.2   Kawabata system

Fabric handle or hand has traditionally been assessed by experts who arrive at an overall judgement on quality after manipulating the fabric with their hands. This system requires years of experience and can obviously be influenced by the personal preferences of the assessor. Professor Kawabata of Japan has carried out a great deal of work with the aim of replacing the subjective assessment of fabrics by experts with an objective machine-based system which will give consistent and reproducible results [16–18]. It is generally agreed that the stimuli leading to the psychological response of fabric handle are entirely determined by the physical and mechanical properties of fabrics. In particular the properties of a fabric that affect its handle are dependent on its behaviour at low loads and extensions and not at the level of load and extension at which fabric failure occurs. It is this region of fabric behaviour that has traditionally been measured and for which specifications have been written.

10.2.1 Subjective assessment of fabric handle

The first part of Kawabata's work was to find agreement among experts on what aspects of handle were important and how each aspect contributed to

Table 10.1 The definitions of primary hand

Hand		Definition
Japanese	English	
Koshi	Stiffness	A stiff feeling from bending property. Springy property promotes this feeling. High-density fabrics made by springy and elastic yarn usually possess this feeling strongly.
Numeri	Smoothness	A mixed feeling come from smooth and soft feeling. The fabric woven from cashmere fibre gives this feeling strongly.
Fukurami	Fullness and softness	A bulky, rich and well-formed feeling. Springy property in compression and the thickness accompanied with warm feeling are closely related with this feeling (*fukurami* means 'swelling').
Shari	Crispness	A feeling of a crisp and rough surface of fabric. This feeling is brought by hard and strongly twisted yarn. This gives a cool feeling. This word means crisp, dry and sharp sound made by rubbing the fabric surface with itself).
Hari	Anti-drape stiffness	Anti-drape stiffness, no matter whether the fabric is springy or not. (This word means 'spread').
Kishimi	Scrooping feeling	Scrooping feeling. A kind of silk fabric possesses this feeling strongly.
Shinayakasa	Flexibility with soft feeling	Soft, flexible and smooth feeling.
Sofutosa	Soft touch	Soft feeling. A mixed feeling of bulky, flexible and smooth feeling.

the overall rating of the fabric. For each category of fabric four or five properties such as stiffness, smoothness and fullness were identified and given the title of primary hand. The original Japanese terms for primary hand together with their approximate English meaning are shown in Table 10.1. These terms demonstrate the difficulty of describing handle and the apparent overlap of some of the terms used.

Primary hand values were rated on a ten point scale where ten is a high value of that property and one is its opposite. The properties that are regarded as primary hand and the values of these that are considered satisfactory differ among fabric categories such as men's summer suiting,

Table 10.2 Primary hands for men's winter suits

Koshi
Numerii
Fukurami

Table 10.3 Primary hands for men's summer suits

Koshi
Fukurami
Shari
Hari

Table 10.4 Primary hands for women's thin dress fabrics

Koshi
Hari
Shari
Fukurami
Kishimi
Shinayakasa

men's winter suiting and ladies' dress fabrics. Some of the properties considered primary for these categories are shown in Tables 10.2–10.4. The primary hand values are combined to give an overall rating for the fabric in its category. This is known as the total hand value and it is rated on a five point scale where five is the best rating. The primary hand values are converted to a total hand value using a translation equation for a particular fabric category which has been determined empirically.

As a result of this work books of fabric samples for each of the primary hands were produced by the Hand Evaluation and Standardisation Committee (HESC) together with standard samples of total hand in each of five categories:

1   Men's winter/autumn suiting.
2   Men's summer suiting for a tropical climate.
3   Ladies' thin dress fabrics.
4   Men's dress shirt fabrics.
5   Knitted fabrics for undershirts.

The purpose of these standards is to act as a reference to help the experts to produce more uniform assessments of fabric handle.

A problem with the system as originally conceived is that of there being a 'best' fabric in each category; that is, a fabric that scored the maximum points for total hand value in a particular category would be universally regarded as the best fabric that could be produced for that end use. It has been found [19] that although this may be true within one country there are differences between countries in their perception of the mix of properties required for a particular end use.

10.2.2 Objective evaluation of fabric handle

The second stage of Kawabata's work was to produce a set of instruments with which to measure the appropriate fabric properties and then to correlate these measurements with the subjective assessment of handle. The aim was that the system would then enable any operator to measure reproducibly the total hand value of a fabric.

The system which was produced is known as the KESF system and consists of four specialised instruments:

FB1	Tensile and shearing
FB2	Bending
FB3	Compression
FB4	Surface friction and variation

These instruments measure the tensile, compression, shear and bending properties of the fabric together with the surface roughness and friction. A total of 16 parameters are measured, all at low levels of force, which are intended to mimic the actual fabric deformations found in use. The quantities measured are listed in Table 10.5.

The properties are measured in the following ways.

The tensile properties are measured by plotting the force extension curve between zero and a maximum force of 500 gf/cm (4.9 N/cm), the recovery curve as the sample is allowed to return to its original length is also plotted to give the pair of curves shown in Fig. 10.23. From these curves the following values are calculated:

Tensile energy WT = the area under the load strain curve
(load increasing)

$$\text{Linearity } LT = \frac{WT}{\text{area triangle } OAB}$$

$$\text{Resilience } RT = \frac{\text{area under load decreasing curve}}{WT} \times 100\%$$

The compressional properties are measured by placing the sample between two plates and increasing the pressure while continuously

Table 10.5 The 16 parameters measured by the Kawabata system describing fabric mechanical and surface properties

Tensile	LT	Linearity of load extension curve
	WT	Tensile energy
	RT	Tensile resilience
Shear	G	Shear rigidity
	2HG	Hysteresis of shear force at 0.5°
	2HG5	Hysteresis of shear force at 5°
Bending	B	Bending rigidity
	2HB	Hysteresis of bending moment
Lateral compression	LC	Linearity of compression thickness curve
	WC	Compressional energy
	RC	Compressional resilience
Surface characteristics	MIU	Coefficient of friction
	MMD	Mean deviation of MIU
	SMD	Geometrical roughness
Fabric construction	W	Fabric weight per unit area
	T_0	Fabric thickness

monitoring the sample thickness up to a maximum pressure of 50gf/cm^2 (0.49N/cm^2). As in the case of the tensile properties the recovery process is also measured. The quantities LC, WC and RC are then calculated in the same manner as LT, WT and RT above.

In order to measure the shear properties a sample of dimensions $5 \text{cm} \times 20 \text{cm}$ is sheared parallel to its long axis keeping a constant tension of 10gf/cm (98.1mN/cm) on the clamp. The following quantities are then measured from the curve as shown in Fig. 10.24:

Shear stiffness G = slope of shear force–shear strain curve
Force hysteresis at shear angle of 0.5° 2HG = hysteresis width of curve at 0.5°
Force hysteresis at shear angle of 5° 2HG5 = hysteresis width of curve at 5°

In order to measure the bending properties of the fabric the sample is bent between the curvatures -2.5 and 2.5cm^{-1} the radius of the bend being 1/curvature as shown in Fig. 10.25. The bending moment required to give this curvature is continuously monitored to give the curve shown in Fig. 10.26. The following quantities are measured from this curve:

Bending rigidity B = slope of the bending moment – curvature curve
Moment of hysteresis 2HB = hysteresis width of the curve

The surface roughness is measured by pulling across the surface a steel wire 0.5 mm in diameter which is bent into a U shape as shown in Fig. 10.27.

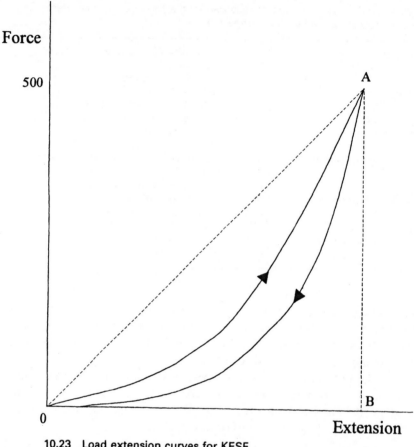

10.23   Load extension curves for KESF.

The contact force that the wire makes with the surface is 10 gf (98.1 mN). A plot of the height variation against distance is shown in Fig. 10.28. The value that is measured is SMD = mean deviation of surface roughness.

The surface friction is measured in a similar way by using a contactor which consists of ten pieces of the same wire as above as is shown in Fig. 10.29. A contact force of 50 gf is used in this case and the force required to pull the fabric past the contactor is measured.

A plot of friction against distance travelled is shown in Fig. 10.30 from which the following values are calculated:

MIU = mean value of coefficient of friction
MMD = mean deviation of coefficient of friction

All these measurements are then converted into primary hand values by a set of translation equations and the total hand values are then

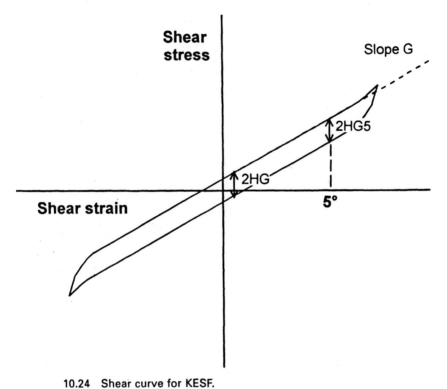

10.24   Shear curve for KESF.

Bending moment

10.25   Forces involved in fabric bending.

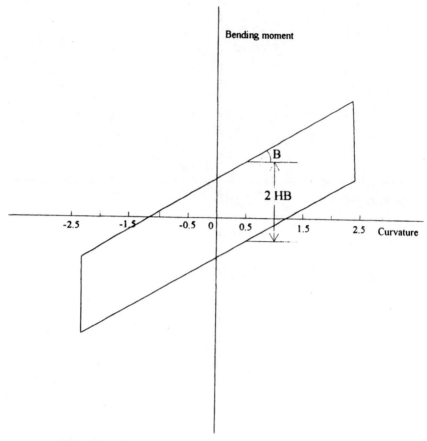

10.26 Plot of bending moment against curvature.

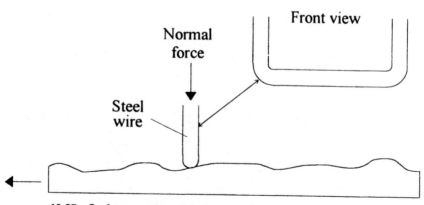

10.27 Surface roughness measurement.

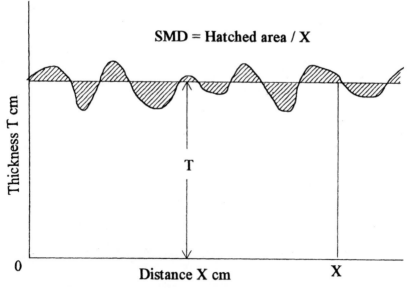

SMD = Hatched area / X

Thickness T cm

T

0

Distance X cm   X

10.28   Surface thickness variation.

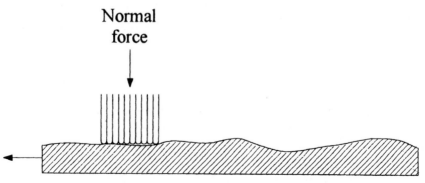

Normal
force

10.29   Surface friction measurement.

calculated from these primary hand values by the use of a second translation equation. Typical results for a summer suiting fabric are shown in
Table 10.6.

The results can also be displayed in the form of a chart as shown diagrammatically in Fig. 10.31. Here the results have been normalised by the
standard deviation of each of the corresponding characteristic values or
hand values using the following relationship:

$$x = \frac{(X - \overline{X})}{\sigma}$$

Table 10.6 Hand values for a summer suiting

Total hand	
THV	3.5
Primary hand	
Koshi	6.1
Shari	6.5
Fukurami	3.5
Hari	6.8

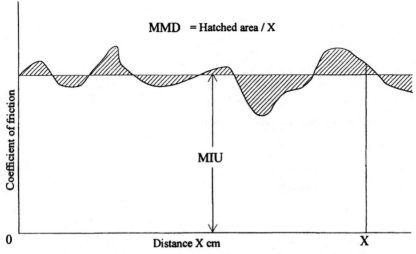

10.30   Surface friction variation. MIU is the mean value of the coefficient of friction.

where: x = normalised mean,
X = measured parameter,
$\bar{X}$ = mean value of property for typical fabric,
σ = standard deviation of property for typical fabric.
By normalising the results they can all be plotted on the same scale. If the values on the chart are joined together a 'snake' chart is produced from which it can be readily seen which fabrics differ from the average. Guidelines can then be drawn on the chart as in Fig. 10.31 showing the good zone into which the parameters of high-quality fabrics fall.

10.3   FAST: Fabric Assurance by Simple Testing

This system is specifically designed by CSIRO for use by tailors and worsted finishers to highlight problems that may be encountered in making a fabric

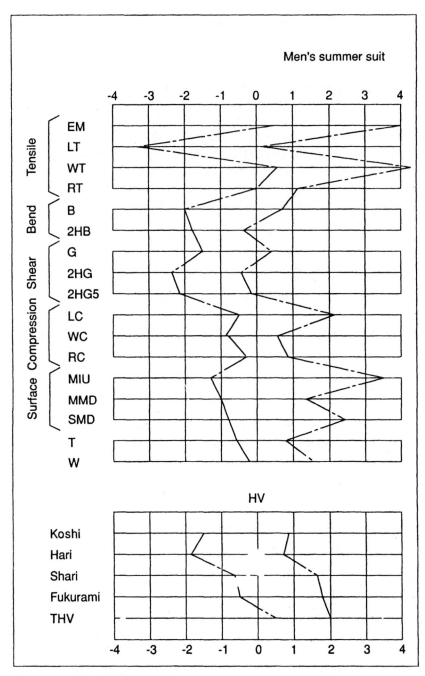

10.31   HESC data chart.

into garments [20]. The system is claimed to be much simpler and more robust than the KESF system.

Most of the parameters measured distinguish between loosely constructed fabrics which readily distort and the more tightly constructed ones which do not distort as easily. The fabrics that are easily distorted can be the source of problems when they are handled because the forces which are involved in cutting them or feeding them into machines can change the shape and size of the fabric. On the other hand firmly constructed fabrics, although they can be handled easily, can give problems in some areas of garment construction where a certain amount of distortion is deliberately introduced into the fabric. Typical areas are moulding and overfeeding of seams where a three-dimensional effect is being produced.

The FAST system comprises four test methods:

FAST 1 Compression meter
FAST 2 Bending meter
FAST 3 Extension meter
FAST 4 Dimensional stability test

The first three methods have purpose-designed instruments whereas the fourth method requires no specialised equipment.

10.3.1 Compression

The fabric thickness is measured on a $10\,cm^2$ area at two different pressures, firstly at $2\,gf/cm^2$ ($19.6\,mN/cm^2$) and then at $100\,gf/cm^2$ ($981\,mN/cm^2$) using the apparatus shown in Fig. 10.32. This gives a measure of the thickness of the surface layer which is defined as the difference between these two values. The fabric is considered to consist of an incompressible core and a compressible surface. The fabric thickness measurements are repeated after steaming on an open Hoffman press for 30s in order to determine the stability of the surface layer.

10.3.2 Bending length

The fabric bending length of the fabric is measured using an automated test method, shown in Fig. 10.33, which is similar to that described in BS 3356 but which uses a strip 5cm wide. The bending rigidity, which is related to the perceived stiffness, is calculated from the bending length and mass/unit area. Fabrics with low bending rigidity may exhibit seam pucker and are prone to problems in cutting out. They are difficult to handle on an automated production line. A fabric with a higher bending rigidity may be more

10.32   FAST compression meter.

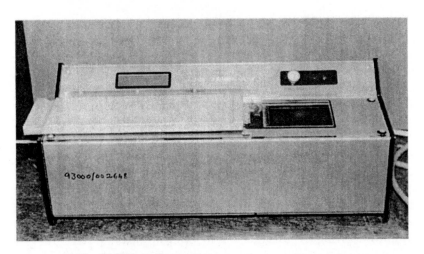

10.33   FAST bending meter.

manageable during sewing, resulting in a flat seam but may cause problems during moulding.

$$\text{Bending rigidity} = 9.8 \times 10^{-6} MC^3 \ (\mu \text{N m})$$

where C is bending length and M is mass per unit area.

10.34   FAST extension meter.

10.3.3 Extensibility

The extension of the fabric is measured in the warp and weft directions at three fixed forces of 5, 20 and 100 gf/cm (49, 196 and 981 mN/cm) (sample size tested 100 mm × 50 mm) using the apparatus shown in Fig. 10.34. The extension is also measured on the bias in both directions but only at a force of 5 gf/cm (49 mN/cm): this enables the shear rigidity to be calculated.

Low values of extension give problems in moulding, produce seam pucker and give difficulties in producing overfed seams. High values of extension give problems in laying up and such fabrics are easily stretched during cutting with a consequent shrinkage to a smaller size afterwards. Problems also occur in cases of high extension with matching checks and patterns owing to the ease with which the material distorts. Low values of shear rigidity have a similar effect to high values of extension as the fabric easily distorts, giving rise to difficulties in laying up, marking and cutting. A high value of shear rigidity means a fabric that is difficult to mould and where there are problems with sleeve insertion.

The formability of the fabric is calculated from the longitudinal compressibility and the bending rigidity. For the purposes of calculation the longitudinal compression modulus is assumed to be equal to the extension modulus. The formability measures the degree of compression in the fabric plane sustainable by it before buckling occurs. Low values of formability indicate a fabric that is likely to pucker when made into a collar or cuff.

F A S T CONTROL CHART FOR TAILORABILITY

FAB.ID   : tw25/75
END USE :   SOURCE:
REMARK :   DATE   :

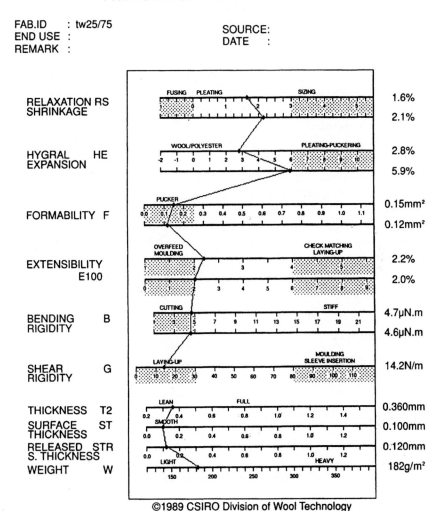

RELAXATION RS SHRINKAGE	1.6%
	2.1%
HYGRAL HE EXPANSION	2.8%
	5.9%
FORMABILITY F	0.15mm²
	0.12mm²
EXTENSIBILITY E100	2.2%
	2.0%
BENDING B RIGIDITY	4.7µN.m
	4.6µN.m
SHEAR G RIGIDITY	14.2N/m
THICKNESS T2	0.360mm
SURFACE ST THICKNESS	0.100mm
RELEASED STR S. THICKNESS	0.120mm
WEIGHT W	182g/m²

©1989 CSIRO Division of Wool Technology

10.35   FAST chart.

10.3.4 Dimensional stability

In order to measure dimensional stability the fabric is dried in an oven at 105°C and measured in both the warp and weft directions to give the length L_1.

It is then soaked in water and measured wet to give the wet relaxed length L_2. It is then redried in the oven and measured again to give the length L_3. The following values for dimensional stability are then calculated from these measurements for both warp and weft:

$$\text{Relaxation shrinkage} = \frac{L_1 - L_3}{L_1} \times 100\%$$

$$\text{Hygral expansion} = \frac{L_2 - L_3}{L_3} \times 100\%$$

High values of shrinkage in a fabric produce problems of garment sizing due to panels shrinking; seam pucker may form in the final pressing stage. A small amount of shrinkage (usually below 1%) is required for fabrics intended to be pleated.

A high value of hygral expansion can lead to loss of appearance in humid conditions as the fabric increases in dimensions under such conditions. The seams can also pucker in these conditions as the sewing thread prevents relative movement of the fabrics.

The whole of the results are plotted on a chart, shown in Fig. 10.35, which is similar to the chart produced by the KESF system (Fig. 10.31). The shaded areas show regions where the fabric properties are likely to cause problems in garment manufacture. These limits have been determined from experience and apply only to the worsted suitings for which the system was originally designed.

General reading

Bishop D P, 'Fabrics: sensory and mechanical properties', *Text Progress*, 1996 **26** (3).

References

1. Peirce F T, 'The handle of cloth as a measurable quantity', *J Text Inst*, 1930 **21** T377.
2. BS 3356 Method for determination of bending length and flexural rigidity of fabrics.
3. BS 5058 Method for the assessment of drape of fabrics.
4. BS EN 22313 Textile fabrics. Determination of the recovery from creasing of a horizontally folded specimen by measuring the angle of recovery.
5. Hilyard N C, Elder H M and Hari P K, 'Time effects and the measurement of crease recovery', *J Text Inst*, 1972 **63** 627.
6. Matsudaira M and Qin H, 'Features and mechanical parameters of a fabric's compressional property', *J Text Inst*, 1995 **86** 476.
7. Hearle J W S, 'Shear and drape of fabrics,' in Hearle J W S, Grosberg P and Backer S, 'Structural mechanics of fibres, yarns, and fabrics', Vol 1, Wiley, Chichester, 1969.
8. Treloar L R G, 'The effect of test-piece dimensions on the behaviour of fabrics in shear', *J Text Inst*, 1965 **56** T533.
9. Cusick G E, 'The resistance of fabrics to shearing forces', *J Text Inst*, 1961 **52** T395.

10. Kothari V K and Tandon S K, 'Shear behaviour of woven fabrics', *Text Res J*, 1989 **59** 142.
11. Kilby W F, 'Planar stress–strain relationships in woven fabrics', *J Text Inst*, 1963 **54** T9.
12. Leaf G A V and Sheta M F, 'The initial shear modulus of plain-woven fabric', *J Text Inst*, 1984 **75** 157.
13. Spivak S M and Treloar L R G, 'The behavior of fabrics in shear', *Text Res J*, 1968 **38** 963.
14. Lindberg J, Waesterberg L and Svenson R, 'Wool fabrics as garment construction materials', *J Text Inst*, 1960 **51** T1475.
15. Carr W W, Posey J E and Tincher W C, 'Frictional characteristics of apparel fabrics', *Text Res J*, 1988 **58** 129–136.
16. Kawabata S, *The Standardisation and Analysis of Hand Evaluation*, 2nd edn., The Textile Machinery Society of Japan, Osaka, Japan, 1980.
17. Kawabata S and Masako N, 'Fabric performance in clothing and clothing manufacture', *J Text Inst*, 1989 **80** 19.
18. Kawabata S and Masako N, 'Objective measurement of fabric mechanical property and quality: its application to textile and clothing manufacturing', *Int J Clothing Sci Tech*, 1991 **3** 7.
19. Behery H M, 'Comparison of fabric hand assessment in the United States and Japan', *Text Res J*, 1986 **56** 227–240.
20. Ly N G, Tester D H, Buckenham P, Roczniok A F, Adriaansen A L, Scaysbrook F and De Jong S, 'Simple instruments for quality control by finishers and tailors'. *Text Res J*. 1991 **61** 402.

11.1 Definitions of quality

The meaning of the term quality is elusive: everybody has their own idea of what is meant by it but it is difficult to express the idea in a concrete form. However, in order to produce a quality product, manufacturers need to have a definition of quality which will allow them to measure how far their products meet the requirements. A number of different definitions of quality have been put forward [1]; each one has its strengths and weaknesses:

1 **Transcendent**. This is the meaning that many people connect with the word quality. It implies that the product has an elusive something that makes the product better than all the competing products. Another view is that the product is superior to all competing products in every way possible. The problem with this approach is that quality cannot be defined in a way that can be used for quality management.

2 **Product based**. In this definition quality is viewed as a quantifiable attribute based on the product's performance in fields such as durability or reliability. Because it is quantifiable then quality can be determined objectively.

3 **User based**. In this definition quality is considered to be an individual matter and the highest quality products are those that best satisfy the customer's preferences. The drawback of this definition is that consumer preferences vary widely so that it is difficult to aggregate these preferences into products that have sufficiently wide appeal.

4 **Manufacturing based**. This definition is concerned with engineering and manufacturing practices based on conformance to requirements or specifications. These specifications are set by design and any deviation from them implies a reduction in quality. Excellence is not necessarily in the eye of the beholder but rather in the standards set by the organisation.

5   **Value based**. Quality in this instance is defined in terms of costs and prices as well as a number of other attributes. The consumer's decision is then based on quality at an acceptable price so that the 'best buy' is not necessarily the cheapest or the one with the highest 'quality' but the one that offers the best combination of the two.

11.2   Types of quality

The actual manufacture of a product is not the only area where quality has to be considered. The parts of the process include the following.

Quality of design

This can be considered as the value inherent in the design. It is a measure of the excellence of the design in relation to the customer's requirements. The production of a quality product starts with its design. The initial meeting of the customer's requirements and the continued functioning of the product throughout its lifetime depend on choice of materials, construction and processes.

Quality of conformance

This is a measure of the fidelity with which the product taken at the point of acceptance conforms to the above design. This is the area that is usually thought of as the province of quality assurance. However, the overall quality depends also on the design as performance cannot be introduced at this stage if it is not present in the original design.

Quality of use

This is a measure of the extent to which the user is able to secure continuity of use from the product. Provided material is being produced which conforms to specification, the length of time the product lasts in use depends on the original design.

Quality of customer service

This is a measure of factors such as the speed of response to orders, the response to customer returns and complaints, the speed and quality of installation and servicing and the initial availability of the product.

11.3   Quality control

The term quality control used by itself has a very narrow meaning and it is generally taken to mean the maintenance of product quality by the regular inspection of critical stages in the manufacturing process. The inspection is carried out on a limited number of items selected as being representative of the current production and the results recorded chronologically on control charts. The appropriate sampling plan is determined by statistical quality control techniques which are also concerned with the monitoring of the sample means and ranges to give warning of the process moving out of control. Quality control is a process for maintaining current standards not for creating new or improved standards.

11.4   Quality assurance

The term quality assurance covers all the processes within a company that contribute to the production of a quality product. It does not just cover the final testing of the product before shipment and it is not solely concerned with testing the product.

11.5   ISO 9000

ISO 9000 is a standard for quality management system. It is different from and additional to any product standard that individual manufactured items have to conform to. The standard has evolved from BS 5750 which in turn had its origins in the needs of the armed forces to procure large quantities of goods from approved suppliers with the utmost reliability. It represents a move away from inspection of the product for conformity to specification towards the requirement as part of the placing of a contract for a proper quality management scheme to be in place. The aim is to do away with inspection of the goods by the purchaser as they arrive and to replace it with periodic checks on the manufacturers to see that they are conforming to the management standard.

The defence standards were rationalised by the British Standards Institute into one set of standards which could be applied to any organisation. The standard is therefore independent of company size, manufacturing methods or product. A quality system is set up that covers the whole company and the system is registered by an inspecting authority which inspects for conformance to the standard. This inspection can be carried out at any time thereafter to ensure that the company is maintaining its standards. A registered firm can then advertise that it conforms to ISO 9000 so that in theory other firms can buy their goods in confidence. In practice many large firms and organisations require that their suppliers are registered so that companies have to register in order to be considered as suppliers.

ISO 9000 is based on the definition of quality which is found in ISO 8402 that is: 'the totality of features and characteristics of a product or service that bear on its ability to satisfy stated or implied needs'.

The ISO 9000 standard itself consists of guidelines for selecting the correct quality system as there are three possible systems for manufacturing industry. The system that is relevant to a particular company depends on how much of the process from design to service is undertaken by that company. The systems for manufacturing are:

- ISO 9001 Quality Systems – Model for quality assurance in design/development, production, installation and servicing. This is the most comprehensive of the three systems and is intended for a company that is specifically required to design the product. The standard also covers the manufacture, installation and servicing of the product.
- ISO 9002 Quality Systems – Model for quality assurance in production and installation. This system is intended for a manufacturer who is producing an established specification or design. This is the most frequently encountered standard and most textile products fall into this category.
- ISO 9003 Quality Systems – Model for quality assurance in final inspection and test. This system is intended for companies whose business is in inspecting and testing products that are supplied to them.

The standard is not directly concerned with the actual properties or design of the product – these will have been decided earlier – but with guaranteeing that the product is always manufactured in the same way, to the same specifications, that no substandard raw material is used in production and that any rejects do not find their way into the output.

The concerns of the standard are really with good organisational practice and it involves complete documentation of the whole process together with internal and external checks to ensure that everything is being run according to these written instructions.

The standard places great stress on writing down all instructions for each stage of the process plus records which identify all work passing through and its status. This material is contained in a quality manual which is the main document for the standard; it contains samples of all documents such as forms that are in use. The system needs to be flexible in order to take account of improved production and testing methods and also to be responsive to customer feedback. One of the ways of doing this is to have controls over the issue and recall of documents so that there is no danger of out-of-date instructions being utilised.

The quality system includes consideration of the following areas.

11.5.1 Enquiries and orders

The quality system begins with the first contact a prospective customer makes with the company. It is suggested that there should be a standard form for recording all enquiries, which are then at some point entered into a central record. A procedure should also be in place so that the appropriate action is taken to follow up any enquiry.

11.5.2 Purchased material or services

The first requirement for incoming material is that a specification should exist for each separate type of material that is purchased. This specification can be either for the supplier's standard product, one that is demanded by the customer or one that is agreed between both parties. Material is only purchased from a list of qualified suppliers who meet one of the following criteria:

- The supplier delivers a product to an agreed national or international standard and there is an agreed system of checking the quality of supplied material in place. This could be either that the customer accepts the supplier's quality assurance system or each shipment is supplied with the appropriate test data or batch or sample testing is carried out by either party.
- There is on-site inspection and test by the customer.
- There is an ISO 9000 system installed by the supplier.

Once the material has entered the company it has to be controlled by strict record-keeping and labelling which includes the inspection status and whether the material has been verified as conforming to standard. There also should be guidelines for the maintenance, storage, handling and use of the material while it is in the manufacturer's possession.

11.5.3 Inspection and testing

Each test used must have its own written procedure together with a statement of the accuracy and suitability of any test equipment, details of the required test environment and personnel needs. The type accuracy and completeness of the data recorded needs to be considered. If, as is usually the case, not every item is tested then a sampling plan needs to be put forward. It is important to consider what purpose is served by the tests and what action is to be taken about any results.

If testing takes place throughout the manufacturing process rather than just on the final product, the inspected status of all material at all stages must be identifiable. Written control procedures are required to show whether material has been inspected and approved. Reject material must be clearly marked as such and written procedures should exist detailing how it should be disposed of. It is not sufficient merely to remove non-conforming material, there must also be a feedback mechanism in place which finds the cause of defective material so that a reoccurrence of the fault is prevented and so that corrective action is promoted which will improve the system. Defects may be caused by incorrect working methods and failure to adhere to instructions. Alternatively the design or product specification may be at fault.

11.5.4 Calibration of test equipment

The test equipment in use must have the necessary accuracy and precision to measure the product to the required degree of accuracy. The equipment must be calibrated prior to use and then recalibrated at regular intervals. The calibration of all measuring equipment, including clocks and rules, must be traceable back to original national and international standards. There must be a statement of maximum allowed intervals between calibration and documentary evidence of the calibration.

11.5.5 Organisational structure

The lines of communication and authority within the company need to be defined, in particular any co-ordination between different activities and the specific quality responsibilities. The standard has to be put in place from the top down and it is considered necessary to have the person who is in overall charge of the quality programme at a suitable level in the company management, preferably at board level.

11.5.6 Quality audit

A quality audit is a check of all the various operations to see whether they conform to the company's own standards. The check can be internal in the first instance but if a company wishes to be registered to ISO 9000 then the audit has to be carried out by an external body licensed to grant the necessary certificate. The audit has also to be carried out at intervals to guarantee continuing conformity.

11.5.7 Training

A requirement of the standard is that staff should be trained to the appropriate level to fulfil their function. Records should be kept of each individual's training which are updated whenever any courses are completed.

11.6   Textile product labelling

There are two types of information conveyed by labels on textile products: some information is optional such as size, manufacturer's identification, country of origin or care instructions; other information, however, has statutory requirements which include fibre content and flammability warning (only on certain garments) [2, 3].

11.6.1 Fibre content

Under the Trade Descriptions Act it is a legal requirement to show the fibre content of clothing and other textile products for retail sale. For these purposes a textile product is defined as one that contains not less than 80% by weight of textile fibres. Where the percentage by weight of a fibre is given it must be accurate to within 3% of the weight of the total. When calculating the weights for percentage composition the standard moisture allowances have to be included.

The fibre content can be marked on either the product or on the packaging. As most clothing is displayed without packing the fibre content is usually marked on a sewn-in label. Small items such as socks or tights are labelled on the package since a sewn-in label would be awkward to position without affecting the product. Fabric sold on rolls need not be labelled directly but the roll itself must be.

Generally the constituent fibres of a textile product should be quoted in decreasing order of percentage content. However, if one component accounts for at least 85% of the total fibre content then the product may be marked either by the name of the main fibre with its percentage by weight or by the name of the main fibre with the words '85% minimum', for example 'cotton 85% minimum' without direct reference to the minor components.

If the product contains two or more fibres none of which accounts for 85% of the total then the names and percentages by weight of the two main fibres should be stated followed by the names of any other constituent fibres in descending order of weight, with or without a figure for their percentage by weight. If the minor component has a special sales appeal such as '5% cashmere' it must not be quoted out of place and its percentage must be given clearly.

Many garments have more than one type of fabric such as linings, interlinings and trimmings. Major linings need to have a separate reference to their fibre content on the label. Interlinings and pads need not be identified as long as they constitute less than 30% by weight of the finished garment. It is usual, however, to identify the fibre content of the filling in quilted and similar articles. When clothing is sold in more than one piece such as a suit,

Table 11.1 Some generic names for man-made fibres [4]

Generic name	Attribute
Cupro	Cellulose fibre from cuprammonium process
Modal	High wet modulus cellulose fibre
Viscose	Cellulose fibre from viscose process
Acetate	Cellulose acetate
Triacetate	Cellulose triacetate
Acrylic	Polyacrylonitrile
Elastane	Elastomeric polyurethane
Elastodiene	Rubber
Modacrylic	Fibre containing between 35% and 85% of polyacrylonitrile
Polyamide or nylon	Contains amide linkages joined to aliphatic units
Polyester	Ester of a diol and terephthalic acid
Polyethylene	
Polypropylene	

one label is sufficient for the whole ensemble. Entries in mail order catalogues must also give the fibre content in their description of textile products. The names used on the label for the fibres must be the appropriate generic names as shown in Table 11.1, not the manufacturers' proprietary ones, though these can be added for example 'Tactel polyamide'. Silk may not be used as a descriptive word for the properties of other fibres that are not from silkworms.

The terms that can be used for wool have restricted meanings: phrases such as 'fleece wool' and 'virgin wool' may only be used if the wool has not been reclaimed. '100%', 'pure' or 'all' wool may only be used when the product is wholly of wool. The regulations concerning wool allow for inadvertent adulteration providing it does not exceed 2% of the weight of the product. In the case of products that have passed through a carding stage, up to 5% impurities are allowed. Purely decorative effects such as metallic yarns are permitted at a level of up to 7% in 'all wool' structures if they are visible and distinct.

11.6.2 Flammability labels

The Nightwear (Safety) Regulations 1985 require that children's nightwear comprising nightdresses and dressing gowns must pass a flame retardancy test laid down in BS 5722. Because of this it does not have to carry a warning label except when a flame retardant chemical has been used. In that case there must be a label stating 'Do not wash at more than 50°C. Check suitability of washing agent'.

\95/ \60/ \40/	Cotton wash (No bar)	A wash tub without a bar indicates that normal (maximum) washing conditions may be used at the appropriate temperature.
\50/ \40/	Synthetics wash (Single bar)	A single bar beneath the wash tub indicates reduced (medium) washing conditions at the appropriate temperature.
\40/	Wool wash (Broken bar)	A broken bar beneath the wash tub indicates much reduced (minimu washing conditions, and is designed specifically for machine washable wool products
(hand wash symbol)	Hand wash only	Do not machine wash

The number in the wash tub shows the most effective wash temperature.

(triangle CL)	Chlorine bleach may be used
◯	May be tumble dried:
◉	with high heat setting
⊙	with low heat setting
(iron, one dot)	Cool iron - Acrylic, Nylon, Polyester
(iron, two dots)	Warm iron - Polyester mixtures, Wool
(iron, three dots)	Hot iron - Cotton, Linen, Viscose
(P)	May be dry cleaned. Other letters and / or a bar beneath the circle will indicate the required process to the dry cleaner.
✕	A cross through any symbol means DO NOT

11.1 Fabric care symbols.

Babies' garments (under three months) do not have to pass BS 5722 but they must carry a label showing whether or not they meet the flammability standard.

If the garment does not pass the relevant test the label must show 'Keep away from fire'. If the garment does pass the test the label must show 'Low flammability to BS 5722'.

11.6.3 Origin markings

It is no longer necessary to label textile products with their country of origin, however, if such a label is used it must be truthful. There is a special case that if a product uses a UK trademark and/or has a UK headquarters address and the garment is made outside the UK then the country of origin must be stated.

Origin means the country in which the goods were manufactured or produced. That is the country in which the components last underwent a substantial change, for example the making up of a garment. However, the addition of minor trimmings or badges does not count as a substantial change. The markings must specify *one* country. Legends such as 'foreign', 'imported', 'made in South Korea and/or Taiwan' are unacceptable.

11.6.4 Care labelling

In the UK, labelling of garments with care instructions is not compulsory but it is a desirable feature. However, once a garment is so labelled it is a legal requirement that the article is fit for the cleansing or finishing method recommended without any adverse effects such as colour loss or shrinkage.

Care labelling is governed in the UK by the Home Laundering Consultative Council (HLCC) which is working towards a reliance on symbols rather than words for a description of the processes in preparation for eventual international harmonisation of labelling. The symbols shown in Fig. 11.1 are: a washtub with temperature for washing, a triangle for chlorine bleaching, an iron with dots in for temperature of ironing, a circle for dry cleaning and a square with a circle inside to represent tumble drying. In all cases a cross through the symbol indicates 'do not'.

References

1. Ross J E, '*Total quality Management*', Kogan Page, 1994.
2. Ford J, 'Labelling: getting the message home', *Textile Month*, 1988 **Feb**. 32.
3. Bryan R, 'Textile law', *Text Horizons*, 1995 **15** 32.
4. ISO 2076 Generic names for man-made fibres.

Appendix: Conversion factors

SI and metric to imperial units	Imperial to SI and metric units
$1 J = 1 N m = 0.1020 kgf m$	$1 oz = 28.349 g$
$1 J = 0.7376 ft lbf$	$1 lb = 0.453,59 kg$
$1 g = 0.035,274,0 oz$	$1 lbf = 4.4482 N$
$1 kg = 2.204,62 lb$	$1 lbf = 0.453,59 kgf$
$1 kgf = 9.8067 N = 2.2046 lbf$	$1 oz/yd^2 = 33.906 g/m^2$
$1 N = 0.101,97 kgf$	$1 lbf/in^2 = 6895 N/m^2 = 6.895 kPa$
$1 kN/m^2 = 1 kPa = 0.145,038 lbf/in^2$	$1 lbf/in^2 = 703.07 kgf/m^2$
$1 Pa = 1 N/m^2$	$1 in = 25.4 mm$
$1 kPa = 0.295,300 in Hg$	$1 ft = 0.3048 m$
$1 kgf/cm^2 = 9.8067 kPa$	$1 yd = 0.9144 m$
$1 kgf/cm^2 = 14.223 lbf/in^2$	$1 yd^2 = 0.83613 m^2$
$1 cm H_2O = 98.0665 Pa$	$1 in^2 = 6.4516 cm^2$
$1 g/m^2 = 0.029,493,5 oz/yd^2$	$1 ft^2 = 929.03 cm^2$
$1 mm = 0.039,37 in$	$1 turn/in = 39.37 turns/m$
$1 cm = 0.393,701 in$	$1 thread/in = 0.3937 threads/cm$
$1 m = 1.0936 yd$	
$1 m^2 = 1.195,99 yd^2$	
$1 decitex = 0.9 denier$	
$1 denier = 1.1111 decitex$	
$1 turn/m = 0.0254 turns/in$	
$1 thread/cm = 2.54 threads/in$	

Index